Firesign

The publisher and the University of California Press Foundation gratefully acknowledge the generous support of the Kenneth Turan and Patricia Williams Endowment Fund in American Film.

Firesign

THE ELECTROMAGNETIC HISTORY OF EVERYTHING AS TOLD ON NINE COMEDY ALBUMS

Jeremy Braddock

UNIVERSITY OF CALIFORNIA PRESS

University of California Press
Oakland, California

© 2024 by Jeremy Braddock

Library of Congress Cataloging-in-Publication Data

Names: Braddock, Jeremy, author.
Title: Firesign : the electromagnetic history of everything as told on nine
    comedy albums / Jeremy Braddock.
Description: Oakland, California : University of California Press, [2024] |
    Includes bibliographical references and index.
Identifiers: LCCN 2024010224 (print) | LCCN 2024010225 (ebook) |
    ISBN 9780520398511 (cloth) | ISBN 9780520398528 (paperback) |
    ISBN 9780520398542 (ebook)
Subjects: LCSH: Firesign Theatre (Performing group)—History—20th
    century. | Comedians—United States—History—20th century. | Sound
    recordings—United States—History—20th century.
Classification: LCC PN2297.F57 B73 2024 (print) | LCC PN2297.F57
    (ebook) | DDC 792.0973—dc23/eng/20240329
LC record available at https://lccn.loc.gov/2024010224
LC ebook record available at https://lccn.loc.gov/2024010225

33 32 31 30 29 28 27 26 25 24
10 9 8 7 6 5 4 3 2 1

For Astrid and Sylvie

Contents

Preface

A postpunk band from Bristol, UK. A 1976 issue of the Marvel comic book *Defenders*. A line in a John Ashbery poem. A *Chicago Tribune* recipe for Zucchini Mushroom Casserole (courtesy of Mrs. George Tirebiter). A psychedelic jazz album on the Impulse! label. Greg Tate's review of the first De La Soul album. A button John Lennon wore to a 1973 press conference. Sixteen tracks from the hip hop auteur Madlib.[1] Each of these is a reference to the Firesign Theatre, a sui generis Los Angeles comedy group that recorded nine virtuosic albums for Columbia Records between 1967 and 1975.

These diverse citations—and it is not an exhaustive list—show how broadly and how deeply the Firesign albums penetrated US (and UK) culture in the 1970s. They are also evidence of how these records generated their own forms of creativity, covert signs of affiliation sent and sought over unknown distances (though I think I would probably rate Madlib over that casserole). And—because they were not the monologues of a solo stand-up personality or recordings of skits but densely multitracked information-heavy forty-minute fictions sometimes called "movies for the mind"—the Firesign Theatre's albums inspired disparate forms of vernacular intellectual culture.

Radical playwright Barbara Garson heard the Firesign Theatre being played at an autoworkers' commune in Lordstown, Ohio, in 1972.[2] David Hidalgo and Louis Pérez formed the band Los Lobos as students at East LA's Garfield High School in an art class that attracted fans of Mexican folk music, rock and roll, Fellini films, and the Firesign Theatre. "I don't want to call us the intellectual group," Pérez remembered, slightly disingenuously.[3] The UCLA course "Electronic Subcultures" devoted seven class periods to the Firesign Theatre in the early 1970s; Lisa Babner's notes from the lectures ran to forty-six typed pages and can now be read at the Library of Congress.

Firesign albums rewarded, and often required, repeated listenings and discussion. They were almost always listened to collectively. They included "everything," from Kent State to Wounded Knee to the *War of the Worlds* panic to the advent of artificial intelligence and early reality TV, but above all, they were about the transformation of society through its media—before, during, and after the time the group was working. The records were also wildly popular, making the *Billboard* charts alongside the popular music of the album era. Oh, and they sounded great when you were stoned.

I am not from the Firesign Theatre generation. My favorite uncle, who was, started giving Firesign records to my brother and me when I was about twelve. Lacking a lot of the context, I still found them as involving as anything in my parents' record collection (*Revolver*, *Sgt. Pepper*, Tom Lehrer, Janice Ian, Queen) or anything I was reading. Their use of recorded sound and punning and/or literary language—the lowbrow and the highbrow jokes—were powerfully formative and probably led me to play music and do my own recordings, study literary modernism, and think critically and creatively about culture. I recently learned that a peer had a nearly identical experience. In high school, I tried to turn on as many of my friends to Firesign as I could. The owner of a local record store, a few years older than I was, would often have a group of kids over to his place, smoking and drinking at the foot of his waterbed (I know, I know), and I would sometimes squeeze in *Don't Crush That Dwarf* after a Pink Floyd record. I had better success a couple of years later with college friends, including a graduate student named Tim Morton, who taught me literary theory.

When I was twenty-two, I bought a used copy of Greil Marcus's *Lipstick Traces* and was amazed by its synthetic riffs on popular music, Dada, and the postwar avant-garde and by its entertainingly speculative historiography. I noticed that Marcus had dedicated his book to the Firesign Theatre and recognized that I had myself been all along listening to Firesign as a kind of "secret history of the twentieth century," which was the book's subtitle. I was living in California then and learned from a friend that Greil Marcus answered his own phone and was often (those were the days) happy to talk to strangers who called him out of the blue. I got his number and called. It turns out that he had contact info for the Firesign Theatre, who had recently reunited, and I went down and met them at a rehearsal in LA.

It was Firesign themselves who connected me to the fans that had continued to plumb the albums' seemingly endless depths, first in their fanzines and then through the new invention of the internet newsgroup (alt.fan.firesign-theatre). Suddenly realizing how much I didn't know, I began to fill in the gaps of my discography, gratefully received recordings of Firesign's radio programs (about which I had been totally ignorant), and got a sense of the huge amount of fan-produced material associated with the group. Finishing this book many years later, I am more appreciative than ever of the fans' labor in documenting the group's complex history.

I am aware that many fans have been waiting a long time for a proper book on the Firesign Theatre. This may not be the book you were expecting. It is not a critical biography, though it contains plenty of the group's history; though I am a lifelong fan, it is not a hagiographic memorial. I had been at enough midnight screenings of *Monty Python and the Holy Grail* to know that I did not want to restage all the jokes, and anyway, the albums are about much more than jokes, as everyone knew. But it also seemed wrong to write a book that was respectfully distanced from its material. I didn't want to exhume the Firesign Theatre only to bury them again. I wanted instead to try reanimating the Firesign Theatre's work for an audience that now listens differently and has its own existential crises with and beyond the continuing explosion of media technologies that we are apparently expected to accept uncritically. This is a work of scholarship, something Firesign fully deserves, but I have tried for my writing also to evoke the experience of listening to the albums, to write in tune with their sensibility, and respond to the multilayered rabbit holes that in turn inspired the diverse experiences and interpretations of their fans.

What did the fans hear in the Firesign Theatre? The examples I began with suggest a very wide range of readings and homages: science fiction, psychedelia, puerile pranks, avant-garde literature, culture jamming, musicality, improvisation, paranoia, social critique, counterhistories, secret affiliations, blissed-out soundscapes. The music writers of the time, for their part, typically understood the Firesign Theatre as inventing a ludic and dark form of media criticism. Very late in the process of writing the book, I discovered something that seemed to embody both of these styles of response.

The first and only issue of the fanzine *Trailing Clouds of Glory* appeared in London in 1974, printed in 11 × 17 tabloid form and comprising collages and articles mostly about the San Francisco music scene. Just before the record reviews at the end is Phil Vellender and Chrissie Toubkin's "Lone Survivors Guide to Firesign Theatre." Its first two pages are a substantial introduction to Firesign written under the intellectual sign of the left-libertarian anticapitalism of the Situationist International (a short list for further reading recommends Guy Debord's *Society of the Spectacle*, Raoul Vaneigem's *Revolution of Everyday Life*, and Situationist pamphlet *Ten Days That Shook the University*).[4] Next is a full-page annotated bibliography of books and films cited by the Firesign albums or evoked for the authors in their listening sessions. It concludes with Hamish Orr's mock-ups of Firesign's fake advertisements ("My husband's a policeman, and you won't believe how dirty he gets my clothes!"). And on page 16 is Toubkin's diagrammatic "Quick Short Circuit Round the Firesign Theatre," a semiserious skeleton key comprising highly memorizable quotations (apothegms and gags) from the Firesign albums, arranged in a pattern schematizing some of their hallmark themes. Four radials spelling the word MEDIA connect the central terms POWER—GRID—GENERATOR—ENERGY—CONTROL to the compass points of RELIGION, MOTOR, FOOD, and OIL; all of these are encircled by THE AMERIKAN SALESMAN (fig. 1). They were honoring the words of the SI's *Ten Days* pamphlet: "if you want to make a social revolution, do it for fun."[5]

The Firesign Theatre thought about media both *speculatively* (what if road signs could talk? what if you could get meals delivered through your

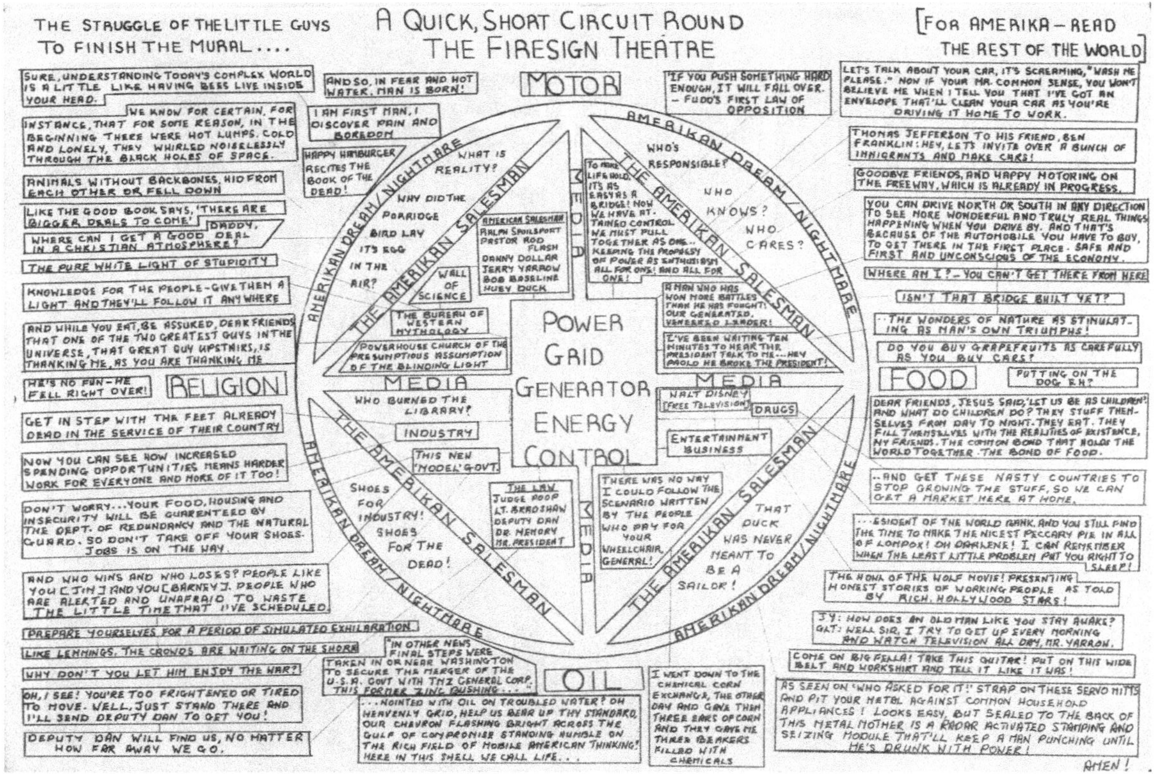

Figure 1. Christina Toubkin's "Quick, Short Circuit Round the Firesign Theatre," an illustration for her five-page article in London's *Trailing Clouds of Glory* fanzine (London 1974), coauthored with Phil Vellender. Courtesy of Phil Vellender.

TV?) and *historically* (how did the history of radio tie the antifascist 1940s to the Vietnam era?). This is why throughout this book I will be referring to the field of "media archaeology" and claiming the Firesign Theatre as media archaeologists *avant la lettre*. To an extent that is hard to imagine today, the Firesign Theatre were able to experiment—not only on their records but in side gigs throughout LA—critically and creatively with cutting edge technologies and with obsolete devices, inventing new techniques and relearning old ones, at the heart of the culture industry and at a time of immense technological and institutional change. The emblem of this work was the iconic CBS Columbia Square studio at the corner of Sunset and Gower and Hollywood, built in 1938 as the La Scala of wartime radio and converted for cutting-edge stereo recording in 1961. (It still stands today, sharing its lot with a twenty-story tower of luxury apartments.)

This book is largely organized chronologically. After a few biographical pages (liner notes), it takes the Columbia albums in order and includes a great deal of the group's history, as well as a general cultural history that at times reaches back to the early twentieth century or forward toward the present. But each of its chapters is also organized around a different medium—album, radio, cinema, artificial intelligence, and television—and discusses how the Firesign Theatre both worked with and critically examined that medium's technology, culture, and history. Each of these mediums incorporated or referred to other mediums: 1970s television was a repository of old movies; the listening practice of psychedelic rock (and taking drugs) collectively evoked the not-distant memory of families gathered around a giant radio in the 1940s; the genealogy of early AI could be traced to the technological spectacle of earlier World's Fairs; the culture of the long-playing record album in many ways resembled the older culture of the book.

Firesign's master medium was of course the album, and they made it a platform for representing all the other media. On 1970's *Don't Crush That Dwarf, Hand Me the Pliers*

    an emergency address from a high school principal
        is heard on a car radio
            by characters in a movie
                that is being broadcast on a television

in the apartment of a character who may have starred in
the movie and is watching
in a city apparently subject to an authoritarian curfew
on the album you are listening to, probably in a dark room
with other people.

The multitrack recording studio became the site for ludic allegories of Marshall McLuhan's "the 'content' of a medium is always another medium" dictum and for much darker allegories about propaganda, surveillance, and social control.[6] This is why their most perceptive critics were usually rock critics (many of them former English majors) for whom the fact that the albums were funny was not always of primary importance. For this reason, I will spend a lot of time discussing how the records were made—in a decade when recording technology transformed rapidly—and somewhat less time describing the group's comedic influences (such as Bob and Ray, Ernie Kovacs, Lord Buckley, the Goons, and *Mad* magazine). As a mediating factor, I will be talking a great deal about the Beatles (dubbed by critic Robert Christgau "the funniest rock stars ever"), with whom the Firesign Theatre had a powerful, and eventually troubled, identification.[7]

Early in the process of writing this book, I was fortunate to participate in a writing group that included a colleague from the Firesign Theatre generation. Trevor Pinch was a pioneering scholar of science and technology, author of an important early book on Robert Moog, and one of the people who invented the field of sound studies.[8] Trevor gave me much-appreciated encouragement at a crucial stage of the project (something I'm sure he did for many people), and I wish he had lived long enough to see this book published. During the year of the cultural acoustics writing group, Trevor was working on a project that the Firesign Theatre would no doubt have been interested in: Stanley Milgram's notorious sound-based experiments investigating the nature of obedience and authority, experiments that began in the shadow of the Eichmann trial in 1961. In the essay draft he shared with us, Trevor was working both to evaluate Milgram's motivations and to understand what it meant to be the hearing subject who participated, believing they were receiving instructions to administer torture. At one point, Trevor made a point of clarifying that "of course we will never listen today as we did then," even though this gap in understanding was one of the things that interested him the most.[9]

I've continued to return to that phrase as I have written a book, a medium that is still relatively healthy, about a series of albums, a medium that barely exists today as it did in 1975. Yet even though the album can seem a bygone form, every Firesign Theatre album as of this writing can be found on a streaming service or YouTube. We are listening more than ever now and not only to music and podcasts. And we still want ways of thinking about our own relation to an even more deeply mediated world— where a massive archive of the world's sounds is available for our listening but in which every second of that listening is measured and quantified— perhaps with a form of utopian pessimism that we can still discover listening anew to the Firesign Theatre.

Liner Notes

Like all liner notes, these pages can be read later or not at all. It is possible to listen to the Firesign Theatre albums without knowing who is speaking, and in that spirit, readers might safely advance to the first chapter. But besides giving a history of things that will come up later, these pages give a sense of conditions that made this unlikely dissident comedy possible, such as a great deal of literary reading. And they begin to tell a story that will be taken up in chapter 1: how the four future members of the Firesign Theatre came together on KPFK in late 1966, were invited to Third Mesa for the solstice, participated in the second-ever performance of MacBird!, helped stage the first Love-In, and were then approached about making a novelty record for the world's biggest label.

David Ossman (Sagittarius, b. 1936) grew up in LA, began studying at Pomona College, and then transferred to Columbia University, where he graduated in 1958. He stayed in New York and became involved in the poetry scene soon to coalesce at St. Mark's Church downtown and got a foot in the door as a replacement announcer on WBAI-FM just as it was joining the nonprofit Pacifica network.[1] By 1960, Ossman was the host of two poetry programs on WBAI, one devoted to readings and the other to conversations on poetics, and in the space of a year, he interviewed more than fifty poets, including Denise Levertov, Robert Creeley, Amiri Baraka, Jackson Mac Low, Allen Ginsberg, Cid Corman, Margaret Randall, Michael McClure, Robert Bly, Paul Blackburn, Jerome Rothenberg, and Tuli Kupferberg (who later formed the Fugs with fellow poet Ed Sanders).[2]

Ossman returned to Los Angeles in 1961 to join the two-year-old Pacifica station KPFK, where he became Drama and Literature Director. Like its sister stations WBAI and Berkeley's KPFA, KPFK specialized in left politics, classical music, folk and blues, literature, public affairs, and children's programming. It avoided popular music. In the next five years, Ossman produced documentaries on Brecht in Hollywood, Parisian Dada, capital punishment, the poetry of Mao Zedong, the Warsaw Ghetto, and many other topics. He programmed readings of Beckett, Baudelaire, Jean Cocteau, and Charles Olson, as well as bilingual readings of the Chilean avant-gardist Vicente Huidobro and Pablo Neruda, and he produced original radio dramas. He also remained active as a poet, teaching poetry at the Free University of California and providing translations for Jack Hirschman's 1965 *Artaud Anthology* and, later, Jerome Rothenberg's groundbreaking ethnopoetic collection *Technicians of the Sacred*. Even after leaving for the "skinny-tie life" at ABC television in spring 1966, Ossman continued to appear on KPFK, hosting the children's poetry program, acting in radio plays, reading *Black Elk Speaks* in half-hour installments, and helping with fundraising.[3] He was also deeply involved with an event, inaugurated in 1963, that would have strong ties to KPFK: the annual "Renaissance Pleasure Faire and May Market." We will return to this in chapter 1. (Faire warning.)

Ossman was succeeded as Drama and Literature Director by Phil Austin (Aries, b. 1941) who, along with his wife Annalee, brought stage acting experience to KPFK. The Austins were particularly interested in the theater of the absurd, a genre that had emerged in the 1950s as a dour reflection on Europe's embrace of fascism and the devastation of two world wars. In the UK, the theatre of the absurd had been popularized by Martin Esslin's BBC radio adaptations, by Grove Press's affordable paperbacks in the US, and by Esslin's hugely influential eponymous study, which defined the field and anointed as its leading figures Samuel Beckett (*Waiting for Godot*) and Eugéne Ionesco (*Rhinocéros*).[4] A propaganda monitor during the Second World War, Esslin saw the new drama as a response to fascism's radical devaluation of language, a phenomenon he found enduring both in the postwar West's mass media advertising and in the coercively totalitarian double-speak of communist states. Though definitive of its era, antecedents could be found, according to Esslin, in commedia dell'arte, Kafka, and silent film comedy.[5] These would all inspire the Firesign Theatre, too, who titled their first album *Waiting for the Electrician or Someone Like Him* after Beckett.

In October 1966, KPFK broadcast a performance of Michel de Ghelderode's protoabsurdist play *Christopher Columbus* (1927), starring Phil and Annalee Austin, David Ossman, and Peter Bergman. A furious burlesque of triumphalist new world "discovery," it concludes incongruously with Buffalo Bill and a huckster resembling P. T. Barnum reading Columbus a celebratory telegram to the tune of Martin Luther's "Ein feste Burg ist unser Gott" and a shower of bullets.[6] The Austins went on to join the Mark Taper Forum's Center Theatre Group, where Phil would appear in John Guare's Vietnam media satire *Muzeeka* alongside future Firesign member Philip Proctor. The Austins also continued to work at KPFK, broadcasting their performance of Ionesco's *The Bald Soprano* in March 1967. That same month, Phil organized a fervent discussion of Austin Black's new volume of Black Arts poetry *The Tornado in My Mouth* with Robert deCoy, Stanley Crouch, and the author. And he oversaw a series of multihour documentaries on the violent colonial history and ominous prophesies of the Hopi Indians, a day of broadcasts that culminated in a twelve-hour "Day of Discussion & Dialogue with Traditional Indian Leaders of the Western United States" attended by hundreds at the First Unitarian Church in downtown LA.[7] The fruit of a project David Ossman had begun in 1962, the Hopi documentaries also would prominently inform *Waiting for the Electrician*.

Austin had grown up in Fresno the child of white proto–New Age artist parents who practiced Vedanta Hinduism. He went to Bowdoin College on scholarship in 1958, then left in 1960 for the Actor's Workshop in San Francisco, a company renowned for its US premieres of Beckett, Genet, Pinter, and Brecht.[8] By 1961, he had moved south to study acting at UCLA, once again leaving before graduating. Austin enlisted in the Army Reserves in 1964, where he received radio training that he would take to KPFK. By the end of 1966, he was Director of Literature and Drama at

KPFK and the engineer of LA's first countercultural overnight program, *Radio Free Oz*.

The host of *Radio Free Oz* was Peter Bergman (Sagittarius, b. 1939). Arriving by motorcycle in July 1966, Bergman announced his LA arrival with a self-arranged midnight screening of a short film, *Flowers*, that he had made the previous year in Berlin. He was then brought to the KPFK studios by Paul Jay Robbins, an activist who had participated in the Selma-to-Montgomery march and was now film critic for the underground *Los Angeles Free Press*, as well as, according to his first wife's memoir, pot dealer for the Byrds.[9] Though KPFK was usually dead air after midnight, Austin and Ossman were then overseeing the station's first ever twenty-four-hour fundraising marathon. Avid response to the late-night riffing (and munificent pledges) revealed an untapped audience to KPFK, which directly asked Robbins and Bergman to commence the station's first overnight program.[10] *Radio Free Oz* quickly became a sensation, inaugurating LA underground radio six months before a new FCC rule would open the FM band to wide national experimentation. It brought rock and roll, rhythm and blues, and Indian ragas to KPFK for the first time, as well as tarot and astrology readings, interviews with scenesters and activists, and, most of all, the participation of LA's night owls and weirdos. Weathering Robbins's departure in October, KPFK's *Radio Free Oz* reached a crescendo on Easter Sunday 1967, when LA's Elysian Park was the site of the first-ever Love-In, an event Bergman had invented and helped coordinate on the air.

Of the four Firesigns, Bergman had taken the most circuitous route to Los Angeles. He had grown up outside Cleveland and matriculated at Yale in 1957. By the beginning of 1961 the FBI had begun to assemble a surveillance file in his name, apparently prompted by Bergman's refusal to sign a loyalty oath for a summer job (no, really) at a small greeting card company in 1960.[11] One informant records his yearslong memberships in the Young People's Socialist League in Ohio and the Congress of Racial Equality in New Haven. The file then goes on to document Bergman's recruitment (by a CIA front, as he would afterward learn) to attend the Soviet-sponsored World Festival of Youth and Students in Helsinki in the summer of 1962, something for which Tom Hayden was also unwittingly auditioning at the very moment he was drafting the Port Huron

Statement.[12] Bergman had by then written a senior thesis on the Industrial Workers of the World and completed a year's postgraduate teaching in labor economics.[13] Anxious to avoid the military conscription threatened by the antisemitic head of his hometown draft board, Bergman then applied for graduate study at the Yale School of Drama. During his year at the Dramat, Bergman directed a production of Clifford Odets's *Waiting for Lefty* and wrote lyrics for a newly discovered Haydn opera titled *House Afire (Die Feuerbrunst)*.

Like Austin, he then enlisted in the Army Reserves, correctly inferring it would be another way of avoiding combat in Vietnam. At basic training in December 1963, Bergman learned he had been nominated for a playwriting fellowship in Berlin sponsored by the Ford Foundation. His benefactor was Benno Frank, a German dramaturg who had worked with Bertolt Brecht and Erwin Piscator before the war and was preparing to direct *House Afire* at Cleveland's renowned interracial theater Karamu House. Having fulfilled his first commitment for the army, Bergman took up residence at the newly founded Literarisches Colloquium in May 1964, ostensibly to work alongside the avant-garde West German writers known as the Gruppe 47. Working under the direction of a British exponent of theatre of the absurd (James Saunders), the cohort of six fellows also included a writer named Tom Stoppard who was then beginning a play with the working title *Rosencrantz and Guildenstern in the Court of King Lear*.[14]

During his Berlin sojourn, Bergman met the filmmakers Bruce Conner, Stan Brakhage, and Shirley Clarke; befriended electronic music pioneer Frederic Rzewski (who would later write an important minimalist piece about the Attica prison uprising); and hung out with several members of the Living Theatre, then developing its pivotal production of *Frankenstein* and performing the anthology piece *Mysteries*, which opened with a participatory dramatization of the plague inspired by Antonin Artaud.[15] Prodigious amounts of hashish were smoked by everyone, Bergman believably recalled, and a great deal else in the case of the Living Theatre. This may or may not have masked the awareness that the Literarisches Colloquium was another cultural Cold War enterprise, a lavish arts initiative housed half a mile from the three-year-old Berlin Wall.[16] Three years later, as the Firesign Theatre were signing their Columbia Records contract, the CIA's involvement with the National Student Association (sponsors of the Helsinki adventure) would be spectacularly revealed in the March 1967 issue of *Ramparts* magazine amid a torrent of disclosures that implicated a host of other liberal cultural organizations like Ford.[17] Firesign responded by closing out the first side of *Waiting for the Electrician* with a satirical allegory of the cultural cold war. It would not be the last time the group would use their recordings to work out the implications of their situation.

After five months in Berlin, Bergman tested the waters writing for *Private Eye* in London before going on to Amsterdam, where he reconnected with friends from the Literarisches Colloquium, the actor (soon to join Living Theatre) Pamela Badyk and cinematographer Gerard Vandenberg. Through them, Bergman met Simon Posthuma and Marijke Koger (who would later form the hippie design collective The Fool) and the poet Simon Vinkenoog (who would collaborate with Bergman on a screenplay for Borges's *Death and the Compass*). And he met Robert Jasper Grootveld.

For several years Grootveld had been the charismatic instigator of a series of theatrical anticommercialist "happenings" in the streets of Amsterdam that would become increasingly political by the spring of 1965. On January 11 of that year, Bergman appeared in the "Stoned in the Streets" happening, which landed him between Grootveld and "pot-art" poet Johnny "the Selfkicker" van Doorn on the cover of the Dutch arts magazine *Ratio*.[18] Heralded by a wall of electric guitars, "Stoned in the Streets" featured a sequence of loose performances—the poetic declamations of van Doorn, a body-painted Koger, Bergman roaming with a tape recorder—all of which led to the revelation of trepanation advocate Bart Huges's "third eye," Grootveld unwinding a long bandage reading "ha ha ha" to show the hole Huges had drilled in his forehead to achieve enlightenment.[19]

Grootveld's happenings became the inspirational foundation of Provo, an anarchist collective inspired by the American civil rights movement, decolonization discourse, peace and disarmament movements, and the international art scene (especially the Dada-inspired Fluxus). Provo earned international renown for its expressive and often very funny engagements, utopian solutions to practical problems (the famous "White Bicycle Plan" of 1965), and theatrical provocations (the March 1966 smoke bombing of

Princess Beatrix's royal wedding to the former Wehrmacht conscript and Hitler Youth Claus von Amsberg). Provo inspired namesake groups in Milan, Brussels, Antwerp, Copenhagen, Stockholm, New York, San Francisco, and Los Angeles before preemptively disbanding in 1967 over concerns about the movement's potential dilution and co-optation.[20] In January 1967, Provo's LA chapter performatively confronted Bergman live on *Radio Free Oz*; led by the composer Joseph Byrd, they took over the KPFK studios, appeared to restrain Bergman, and staged dozens of continuous sign-offs with the national anthem playing repeatedly. In April, Byrd, now leading the avant-garde psychedelic band the United States of America, invited the newly formed Firesign Theatre to perform with them at UCLA's Experimental Arts Festival.

The final Firesign to arrive in LA was Philip Proctor (Leo, b. 1940). A year behind Bergman at Yale, Proctor had starred in two of Bergman's productions and was a member of the Yale Russian Chorus, which toured the Soviet Union in 1959 and performed at the Vienna Communist Youth Festival three years before Bergman attended its follow-up in Helsinki.[21] Proctor moved to New York after graduation, acting in soap operas and understudying Rolfe the singing Nazi in *The Sound of Music*, before landing the lead in an off-Broadway musical titled *The Amorous Flea*, for which he received a *Theatre World* award in 1964. Proctor was back on Broadway the following year understudying in the painfully titled *A Race of Hairy Men! Hairy Men* closed after just four performances but not before Proctor formed a consequential friendship with its star, Brandon de Wilde. Famous for his Oscar-nominated performance in *Shane* (1953), the erstwhile "king of the child actors" had by 1965 become one of the country's leading potheads and was testing the waters as a musician. He and Proctor stayed in New York for another year, occasionally jamming with their friend Gram Parsons as unofficial members of the International Submarine Band.[22] Then, in the fall of 1966, de Wilde persuaded Proctor to join him on a trip back to the West Coast.

On November 26, they accompanied de Wilde's friends Peter Fonda and David Crosby to the mass protests then entering a third weekend on Hollywood's Sunset Boulevard. What became known as the Sunset Strip Riots originated in a commercial dispute. The local Chamber of Commerce, representing up-market restaurateurs, sought to restrict the incursion of grungier youth venues like Whisky a Go Go, the Fifth Estate, and Pandora's Box, which had made the Strip the center of the American rock music world by booking acts like the Doors, the Mothers of Invention, the Byrds, Otis Redding, and Love. Mass arrests followed the imposition of a curfew in July. They grew more intense after Free Speech Movement leader Mario Savio was denied readmission to UC Berkeley on November 7 and, on the following day, Ronald Reagan (who had campaigned against the students) narrowly won a first term as governor of California.[23] November 12 saw the circulation of fliers calling for organized demonstrations starting at Pandora's Box, which was now scheduled for demolition. The notoriously reactionary LA police were happy to provoke a confrontation.

The geographer Mike Davis, who participated as a young demonstrator and organizer for SDS, would pointedly remember the protests as "happenings."[24] The *Los Angeles Free Press* (familiarly, the "Freep") reported that de Wilde and Fonda had come to the Strip with a camera and were researching "a proposed film," later to be known as *Easy Rider*.[25] Both were among the first arrested on the weekend of the twenty-sixth. Proctor, by his own account, avoided arrest by displaying a press pass from the *East Village Other*, an underground New York paper that listed him on the masthead. He then sat on the sidewalk on top of a current issue of the *Freep*, opened to page 3. There, serendipitously above an article on the Amsterdam Provos, Proctor found a photograph of "KPFK newsman Peter Bergman" pointing a microphone at two marines as they arrested a protestor on the Strip the previous weekend.[26] Proctor soon sought out Bergman at the KPFK studios and was invited on the air.

By the end of the month, Proctor, Bergman, Ossman, and Austin had appeared together for the first time on *Radio Free Oz*. They would continue coming together there informally, with conversations and riffs through which they developed the improvisational sociability audible on their albums. In the next four months, meanwhile, they would accept an invitation to attend the Hopi Indians' Soyal ceremony on Third Mesa in Arizona, stage a New Year's Eve performance of Barbara Garson's brand-new satire *MacBird!* live on KPFK, host the Day of Discussion & Dialogue with Traditional Indian Leaders at the First Unitarian Church, stage another fundraising marathon for KPFK, and organize the Elysian Park Love-In.

None of this obviously explains how, days after the Love-In, they would be invited to record an album for the world's largest record label. Nor does it provide an obvious genealogy for the Firesign Theatre as proleptic media archaeologists. But it does give a genealogy for them as what musicologist David McCarthy has named "middle class observers [who] were among the most stringent critics of [middle class] narrowness" imagining forms of involvement in a "society felt to verge on both the utopia of the professionals and apocalyptic race war."[27] And it shows how that dissident creativity, harder to appreciate today, had a powerful source in literary writing. That literary background explains how they, more than any other recording act of the 1960s, could appreciate intuitively the way the media archaeology of the long-playing record album extended back to the history of the book. That story, as well as what happened with Firesign after Proctor arrived at KPFK, is the subject of chapter 1.

1 ALBUM / Talking Book

Waiting for the Electrician or Someone Like Him (1968)

HOW DID THE LP BECOME A BOOK? (ARCHAEOLOGY OF CONTEMPTIBLE MEDIA)

"In the entire history of recorded sound there has never been a more contemptible genre than the comedy album."[1] This statement would have been a fair reaction to the stack of records featured in the August 18, 1968, *New York Times*: besides the eighth albums by (respectively) Jonathan Winters and Bill Cosby, the *Times* surveyed (among others) the celebrity impressions of *Rich Little's Broadway*, the borscht-belt innuendo of Lee Tully's *Join the Love-In, The Very Funny Side of Pete Barbutti*, and *The Slightly Irreverent Comedy of Ron Carey*.[2]

In fact, this contemptuous statement is the first sentence of an article about the Firesign Theatre that appeared in the legendary Michigan zine *Motorbooty* in 1992. An icon of the post-hard-core punk era, *Motorbooty* specialized in freak culture reclamation projects and routinely made a show of taking minority positions. The same 1992 issue also featured (and this will be relevant later) a piece celebrating several of Miles Davis's electric period records, which were then still unfashionable, controversial, and mostly out of print. So it is not surprising that the Firesign Theatre—

whose debut was also incongruously noted in the *Times* roundup—is the exception to *Motorbooty*'s Sick Burn: "Comedy albums have been around since the Edison Canister, yet in all that time, no one has fully explored the comedic possibilities of the 'big 12-inch record.' No one, that is, except the Firesign Theatre."[3]

David Merline's *Motorbooty* essay remains one of the best things ever written about the Firesign Theatre, and it is notable that it appeared seventeen years after the group's last album for Columbia Records, fifteen years after cable television displaced the long-playing record album as comedy's signature medium, a decade after the concept of *album* had itself been transformed by the single-sided eighty-minute digital compact disc.[4] Merline was right to identify comedy's place in recorded sound's long history and he may have also known that he was distilling arguments from an earlier moment of media transition, the late 1960s, when the twelve-inch LP displaced the seven-inch single as youth culture's coin of the realm.

Led by album sales (and their favorable profit margin), Recording Industry Association of America (RIAA) revenues went up by more than $1 billion between 1966 and 1972.[5] During this time of expansion, it became common for the LP's myriad genres, including comedy, to be reviewed by a new kind of cultural arbiter: the professional rock critic. By 1974, that field's self-styled dean, Robert Christgau, would muse about the "strange kind of investment" represented by (for instance) a Bill Cosby album: "You buy it and bring it home and play it and laugh, but then what? Who really plays it again?"[6] The same sentiment was issued by *Creem* editor Robert Duncan, who began a Richard Pryor review on a downer: "It's horrible to review comedy records. After a week of intensive immersion in even the funniest human on earth . . . you're not likely to see much humor."[7]

Despite a decade of chart success for figures like Cosby and Bob Newhart, Christgau and Duncan both understood performed comedy to be misallied with the album form. But they also saw that the field had begun to internalize the problem. In his annual survey, Christgau recognized the Firesign Theatre as the virtuosic inventors of a comedy that was distinct in content ("it avoids gags") and in form ("it is molded to the phonograph record") and that exhibited "structure and dimension that rewards repeated listening" (which was implicitly a feature of album culture in general).[8] For his part, Duncan reviewed stand-up comics like Pryor, George Carlin, and Lily Tomlin, before concluding ambivalently that "the ephemerality of comedy is something the 'new' comedians are trying to eliminate by ... working towards something along the line of spoken literature."

What if these two critics were really saying the same thing, that to mold comedy to the form of the phonograph record was also to make comedy *literary*? And what if this observation has implications that extend much further than the apparently niche genre of comedy into the broader cultures of music and literature, and into their mediums? Questions like these are at the core of this book, which studies the nine albums recorded by the Firesign Theatre in Los Angeles for Columbia Records between 1967 and 1975. Years later, Christgau, unconsciously citing Duncan, called Pryor's albums "a great body of recorded literature."[9] But no one today would think to call Dave Chappelle or Sarah Silverman or Sacha Baron Cohen literary. The fact that such an idea was communicable circa 1974 is not simply a sign of literature's former centrality to popular culture or testimony to the erstwhile international hegemony of records. It also has to do with properties of recorded sound, with practices used for recording and for listening to the vinyl LP in the late 1960s (inscription and playback), and with the format's longer history, all of which are related to the medium of the book in manifold ways (though the history of this association will be surprising).

It is not an accident that *album* was originally, and remains, a word for a kind of book. Why did it take a stoner comedy act, working in the heart of the culture industry at the peak of the counterculture period, to explore this connection most thoroughly? How did they use this insight as the basis of a much wider-ranging investigation of media, culture, and social life? And how did their sui generis novelty albums find a place on the *Billboard* charts alongside the rock, rhythm and blues, soul, and country records of their day, inducing thousands of fans to memorize the albums, later inspiring the most recondite DJs in hip hop to sample them (almost, we might say, bibliographically)?

To answer these questions, I want to begin by thinking about the medium of the *album*. At first glance, the word may seem to refer specifically to the two-sided, twelve-inch vinyl disc. But as the media scholar

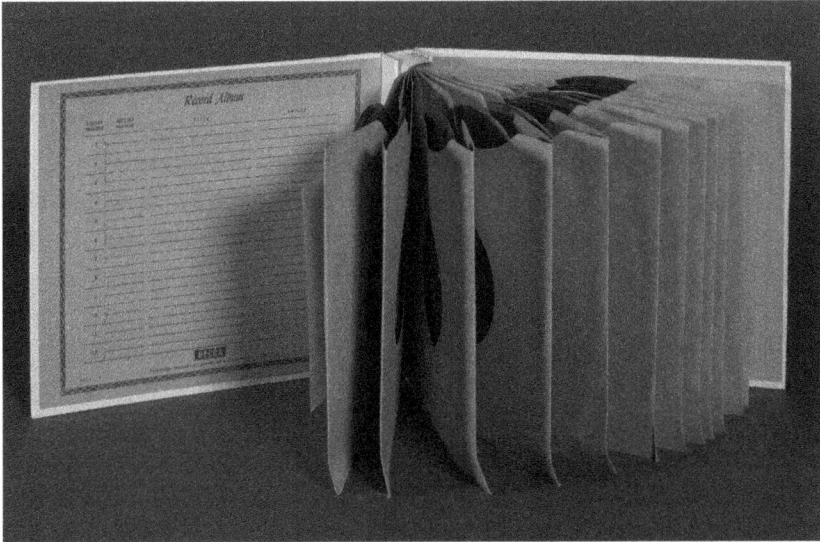

Figure 2. A generic "record album" collecting 78 rpm discs originally purchased in paper sleeves. Record companies would sell similar albums, with cover art, for sets of discs of symphonies or collections of songs. Introduced in 1948, the 33⅓ rpm long-playing record would anachronistically preserve the word *album.* Courtesy of Cornell University Library.

Jonathan Sterne has insisted, when we speak of mediums, or *media*, we are never talking about a single technical device but rather constellations or assemblages "where technologies, institutions, and practices come together in recognizable and repeatable form."[10] This conclusion (and its obverse) is something the Firesign Theatre realized heuristically in their work at CBS's Columbia Square Studio in Hollywood and in side gigs around a rapidly transforming Los Angeles.

"Record album" did not originally refer to a vinyl disc, much less an abstract concept of a musical work, but to a bound booklet of paper sleeves made to hold a series of ten- or twelve-inch 78 rpm shellac discs (fig. 2), which, because of their very short (three-to-five-minute) running time, were often sold in sets, while the "album" itself was often sold separately.[11] But the term outlived the 78 rpm disc, and the material association with a bound book, when it became attached to the new high-fidelity long-playing microgroove vinyl record, introduced by Columbia Records in 1948. By the late 1960s, the term had become still more abstract and conceptual (a sense that would later be extended to compact discs, digital downloads, and streaming). One "recorded an album" in the same sense that I am at this moment "writing a book" on my computer, a year or more before you are holding it in your hands (or reading a pdf scan, or whatever).

A Sternean approach would stress this genealogy, but it would also point out that to speak of the record album in 1967 was not only to speak of the two-sided forty-six-minute vinyl disc (plus cardboard jacket) and its artist but to discuss record companies and their studios, which were staffed by union-trained engineers and musicians and contained an expanding array of technical devices such as microphones, magnetic tape, recorders and mixing boards, filters and effects, and, of course, instruments. It was also to invoke recording and performance conventions, as well as experimental practices (some of which would become regularized), that would be increasingly tied to the dominance of stereophonic sound and the expansion of multitrack recording.[12]

And, finally, the meaning of the LP as a medium was shaped by the act of playing the records, which was itself an affair of media technology (high-fidelity turntables, styli, tone arms, amplifiers, speakers, and so on) combined with habits and techniques of reception (the practices of the audiences who listened). Though headphones slowly became more prevalent, through the early 1970s, listening was often a collective experience, one writer suggesting that album culture "transplanted ritual from temple or theatre into any place where two or three may gather together."[13] This was also true for drugs (an emerita colleague once reminded me), and (eureka) the two experiences usually went well together. As the great Ellen Willis described Van Dyke Parks's 1968 album of multitracked psychedelic pop, "marijuana is to *Song Cycle* what Polaroid glasses were to 3-D movies."[14]

These insights should cue us not only to the increasingly *compositional* dimension of multitrack recording—music that could only have been made in the studio and made no pretense of simulating live performance—but also to the often *theatrical* quality of the music that seemed to solicit the experience of listening in a group (the ritual theater, the 3-D movie). Hence the crowd hubbub and tuning orchestra at the start of *Sgt. Pepper's Lonely Hearts Club Band* (recorded on four-track machines in

1967) and the machinic relocations on Pink Floyd's *Wish You Were Here* (recorded on twenty-four tracks in 1975) through the worlds embedded in sample-based hip hop of the late 1980s, such as when Ice Cube and Big Daddy Kane's phone call to Chuck D. and Flavor Flav in Public Enemy's "Burn Hollywood Burn" results in all four rappers going to the movies (and then back to Kane's place for *Black Caesar* on VHS), or in the skits on *Three Feet High and Rising* by De La Soul, the "bedazzling bookish types" seen by the great critic Greg Tate as following the tradition of Parliament-Funkadelic, Sly and the Family Stone, and the Firesign Theatre.[15]

Firesign Theatre realized this theatricality still more literally, and it should not be surprising to learn that their records were almost always listened to collectively. Assuming its readers would recognize the scene, *Rolling Stone* began its 1971 Firesign profile with the vignette: "Through a Florida garden, stoned, maybe a year ago, to a little house where some friends are. Someone opens the door, says 'shhh.' I sit, mildly irritated. The pipe comes around. They are listening to something. . . ."[16] That private scene became quasi-institutionalized at Cornell University, where the student union hosted monthly four-hour listening parties—"Free Popcorn, Free Firesign & Free People"—starting in 1970.[17] Such stories appear again and again in the mimeographed fanzines of the 1970s and 1980s, on the alt.comedy.firesign-thtre Usenet group that followed them, and in my interviews with fans from the period.

As these all suggest, recorded comedy became literary when it realized the theatricality implicit in psychedelia's employment of asynchronous multitrack recording. To forecast this chapter's argument: the Firesign Theatre may have set out to do something they understood to be a version of radio drama, using multitrack recording to extend further the "dramas of space and time" that radio historian Neil Verma has found to define the medium in the late-1930s US.[18] In practice, however, the culture of the long-playing album was equally, if not more, redolent of the culture of the book. Albums were portable, a small cultural investment, could be collected in a "library," could be listened to again and again in the privacy of domestic space, and could be memorized and otherwise internalized. As we will see, classic histories of reading line up almost exactly with more recent critiques of high-fidelity album culture. This connection was in fact playfully anticipated on rock records like *Ogden's Nut Gone*

Flake by the Small Faces (1968), which recruited "Professor" Stanley Unwin to parody a children's storybook album throughout its second side.

"Molded to the phonograph record" as they were, the Firesign Theatre's ludic dramas were book-like in another way. In the 1930s, the Russian literary theorist Mikhail Bakhtin described the novel as a form that was essentially *comedic*, even if the comedy was often a dissenting and gruesome "world upside down," as was the case in Dostoevsky's novels. Such a précis would neatly summarize the Firesign Theatre, too, as many rock critics understood intuitively: in the first *Rolling Stone Record Guide* (1979), Greil Marcus called the group "horrifying, death-dealing, life-enhancing."[19] But the resonance with Bakhtin goes further than this.

Opposed to the supposedly *monologic* nature of poetry (an adjective internal to stand-up comedy), Bakhtin found the language of the novel to be definitively *heteroglossic*. It is "a living mix of varied and opposing voices, developing and renewing itself" in parodic resemblance of the situation of social life.[20] Of all the literary genres, the novel alone is capable of "reflecting in all its fullness the . . . multiple voices of a given culture, people and epoch" (60). As the only literary genre to emerge after the invention of writing, it is also uniquely capable of incorporating and representing all the other genres (anticipating what sample-based hip hop, which emerged after the apotheosis of the twelve-inch record, would achieve for music). The novel is "a new and large multi-genred genre" attuned to representing social heterogeneity and division; it is "inseparable from social and ideological struggle" (65–68). All of this aptly describes the Firesign Theatre, too, most of all their modernist practice of "doing the police in different voices," which could at times involve appropriating racial and gender mimicry, as well as satires of myriad forms of whiteness—which was in turn one reason they were so attractive to the DJs who sampled them, from Mark the 45 King to Madlib. Bakhtin would further make clear that the heteroglossia of the novel contained both spoken dialects and professional discourses, "tendentious languages, languages of the authorities, . . . languages that serve the specific sociopolitical purposes of the day," as well as, crucially, *media forms* such as newspaper writing and penny dreadfuls (263). This was the point made by the eminent critic Lester Bangs, who described Firesign (in an otherwise savage kill-yr-idols critique) as "masters of Voice" who "have indeed changed our perception of the voices on the airwaves."[21]

Bakhtin's word for the way all these voices are arrayed, in the novel as in social life, is *stratification*, and this term would have a special resonance in the context of the intense racial segregation of late 1960s Los Angeles. On radio programs, the Firesign Theatre would at times perform stereotyped Black voices that directly recalled the practices of minstrelsy or took on Latinx voices (at times speaking Spanish), as with an improvisation that called out the pathetic PR campaigns of the LA police after their murder of the journalist Ruben Salazar in 1970 (the *Dear Friends* LP's "Deputy Dan Has No Friends"). These vocal practices are not defensible and should not be considered available forms of entertainment or critique. But it is possible to condemn them while also observing that every instance of aural blackface on the Firesign Theatre's deliberately scripted albums is intended as a critique of segregation and racism (as I will discuss in later chapters).[22] In this way, the albums can be heard to attempt a politics that did not reproduce the silence that defined so much of LA's white music counterculture and refused the later nihilism of the *National Lampoon* and condescending Black voice parodies of many others, even as they also exemplify the practice Greg Tate resonantly encapsulated in the phrase "everything but the burden."[23]

In the language of Bakhtin's original text, the word *rassloenie* principally signifies the stratification of social classes and groups. But in its English translation, *stratification*—a word with meanings both *archaeological* and *social*—usefully also conveys Bakhtin's emphasis on the history of parodic genres and critical social practices that predate the novel but help to shape its meaning, such as the medieval carnival and early modern commedia dell'arte (a point to which I will return) (68–83).[24] The richness of the translation is important because *stratification* so clearly describes the sound of the Firesign Theatre's multilayered, hypermediated, polyvocal records and also because it suggests a way literary study might intersect and mutually inspire the comparatively recent field of media archaeology, which examines how the past is embedded in obsolete technical devices and how subjectivity evolves in relation to media, composes counterhistories and counterfactual histories, and has been called (because of its emphasis on "doing" media studies) "both a method and an aesthetics of practicing media criticism."[25]

The Firesign Theatre, I suggest, were both virtuosic collective authors of a Bakhtinian heteroglossia and media archaeologists *avant la lettre*, discovering (as this chapter shows) the book that ghosts the heart of the LP album. As I will argue, their second-generation work in the historic birthplace of the still-expanding and renewing Adornian culture industry involved working directly with technologies both obsolete and cutting-edge, relearning older media practices and inventing new ones, devising stories from recovered histories and speculative futures. Of all the artists then recording at CBS Columbia Square in Hollywood, the Firesign Theatre understood best of all what it meant to make futuristic records in the context of Vietnam and the setbacks of 1968 in a studio that had been built to broadcast antifascist radio propaganda during World War II.

WHO WERE THE READERS OF RECORDS?

If at this point this all seems willfully mind-melting, it will at least not be controversial to say that the nine densely plotted, polyvocal albums made by the Firesign Theatre for Columbia Records between 1967 and 1975— between the years of *Sgt. Pepper* and *Wish You Were Here, Are You Experienced* and *Mothership Connection, Smiley Smile* and *The Hissing of Summer Lawns, The Society of the Spectacle* and *Discipline and Punish*— were the most ambitiously and self-consciously literary albums ever to appear on the *Billboard* charts (though they were many other things, too). That ambition is expressed openly on their first record, *Waiting for the Electrician or Someone Like Him*, which explicitly quotes Allen Ginsberg and William S. Burroughs and overtly refers to Beckett (*Waiting for Godot*), Camus (*The Plague*), Artaud (*The Theater and Its Double*), Václav Havel (*The Memorandum*), and Kafka (*oeuvres completes*). The record's first half, moreover, comprises a suite of linked sketches that, while clearly riffing on a famous comedy album from 1961, can also be heard as a travesty of the "Great American Novel" shibboleth, preceding and maybe inspiring US novelists' repeated, vexed engagement with the concept throughout the 1970s.[26] Side 1 of *Waiting for the Electrician* begins with a mordant history of US settler colonialism, follows it with a satire of an authoritarian counterculture "in which Big Brother is the Zig Zag Man and everyone is groovy" (Merline), and concludes by uniting these conceits in the Strangelovian image of "the United States of Being" dropping "eight million hardbound copies of . . . *Naked Lunch*" on Nigeria from a B-29 bomber called the Enola McLuhan. "God, what an awesome moment," intones the pilot as a nuclear blast sounds the successful deployment of the avant-garde literature, "the last pocket of Un-Hip resistance is *out-of-sight!*" (How about that, Ford Foundation.)

Thus, even as their subsequent records would be far more sonically adventurous—enabled by developments in recording technology, their growing expertise in the studio, and the increasing comprehension of Columbia's engineers—Firesign Theatre's literary bona fides were at the beginning (maybe all too) prominently displayed. Yet it is at least as important to note that the Firesign Theatre were *by far* the most frequently and vatically reviewed comedy act in such underground and rock music publications as *Oz, International Times, Rock, Friends, Fusion, Rolling Stone, Crawdaddy!*, the *Berkeley Barb*, the Chicago *Seed*, and the hometown *Los Angeles Free Press*. In *Creem* alone—"America's only rock 'n' roll magazine"—the group was the subject of a cover story, four standalone articles (fig. 3), regular news items, and extravagant reviews like Dave Marsh's frantic panegyric in 1971: "Most frightening of all are the levels at which they apprehend our culture, our technology and ourselves. . . . None of us have the capacity to comprehend everything here in one or two or three sittings."[27] Firesign's David Ossman would immodestly remember *Waiting for the Electrician* as "the most important literary achievement of the alternative movement of the 1960s," but this could never have been the case if the Firesign Theatre had not found an audience that was prepared to recognize a long-playing record album as literature.[28]

In his classic 1987 book *The Recording Angel*, Evan Eisenberg makes provocative allusions between the world of phonograph records and the world of books, an idea the musicologist Arved Ashby approaches more systematically in a study that takes "recordings as a chapter within the history of writing" (the word *phonograph* literally means "written sound").[29] The media archaeology of the album, Ashby might say, includes the book. That insight was a kind of practical knowledge in a society where

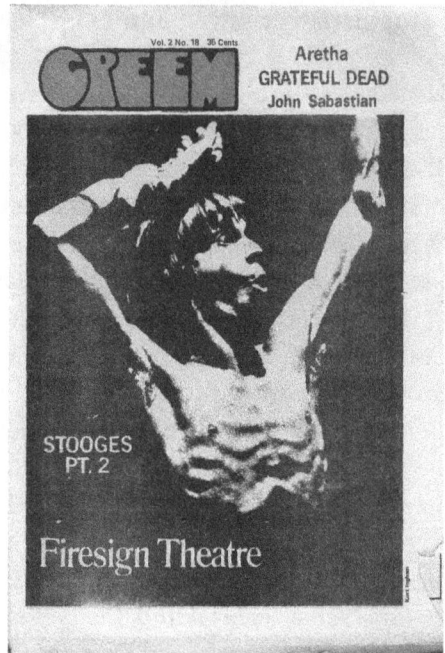

Figure 3. Cover of the December 1970 issue of *Creem*, featuring Iggy Pop and the Firesign Theatre. Photo by Kurt Ingham. Collection of the author.

the number of college English majors tripled during the very years—1950 to 1972—that the high-fidelity long-playing record album came to dominate popular music culture and the recording industry.[30] The same issue of *Fusion* that extolled Firesign as "humor structured like music" also heralded "the delivery of a new manuscript from The Velvet Underground" in its four-column review of *Loaded* (1970).[31]

As Dave Marsh, among many others, would recognize, the Firesign Theatre came to fashion the long-playing record album as a medium for representing and evaluating *all the other media*. Bakhtin might have called this their most "literary" achievement. To approach the question from the beginning of the Firesign Theatre's recording career is to examine a literary field already critically engaged with what has been called a "crisis of textuality" that inspired the audiotape experiments of Ginsberg, Burroughs, Kerouac, and Warhol; marked by a decade of "spoken word" albums from labels like Caedmon, Folkways, and Spoken Arts; and challenged and extended by an emergent field of quasi-theatrical, often spontaneous or public, collective practices by groups like Fluxus, the Living Theatre, Provo, the Diggers, and the San Francisco Mime Troupe (who were all inspired by Antonin Artaud's edict against the printed text).[32] And it is also to name a set of media practices—even media ontologies— that Stevie Wonder would knowingly figure in the title of his 1972 album *Talking Book*. Tracing the genealogy of this oeuvre through literary practices increasingly sublimated as media practices, abetted by audio-technological advances and the Firesign Theatre's increasing facility with studio recording, will reveal a broader story about "recording as a chapter in the history of writing."

HAPPENINGS, DECEMBER 1966

Peter Bergman's *Radio Free Oz* had been the underground phenomenon of 1966, since its mid-July debut on LA's nonprofit Pacifica station KPFK. Engineering the show, Phil Austin often engaged Bergman from behind the glass, and David Ossman "was encouraged to drop by and, you know, be groovy."[33] After Phil Proctor showed up in November, the four future Firesigns quickly discovered a rapport that allowed them to improvise together in more or less absurd ways, bringing listeners along as collaborators or dupes. They staged an ersatz film festival for their first joint appearance on *Radio Free Oz*, commenting on imaginary movies over the piped-in sound of a whirring projector, and then performatively censored the films, to the great offense of the call-in audience. In these early irregular and informal performances, the group began to establish a mode of working together that in time became both an improvisatory end in itself and a workshop for ideas they would revise, script, and record.

Traveling through Turkey and Lebanon in 1965, Bergman had developed an interest in esoteric spirituality, which he now took to the air in his semiserious persona as the late-night "Wizard of Oz." Discovering his astrological affiliation with Austin, Ossman, and Proctor, Bergman began referring to the group collectively as the Oz Firesign Theatre. The music director at KPFK, William Malloch, would remember Peter Bergman as "the spirit of the thing; he was the brain, he was the central nervous system"; David Ossman was the "scribe [who] tried to catch what they came up [with] in improvisation"; Phil Austin "changed from a very straight guy to somebody who was really out there"; and Proctor was "a pro, always, a . . . comedian who could do all kinds of things." "You can always improvise well if you're thoroughly prepared," Malloch went on, "but their preparation was a matter of an ongoing temperament."[34]

The late-night riffs were only one of a range of heterogeneous projects in which they engaged through the end of the year. Bergman had been using the KPFK facilities to prepare short tape collages to use as the opening for each *Radio Free Oz* broadcast; in mid-December, the group used the equipment to record a piece scripted for all four voices (the somewhat puerile "International Youth on Parade"). By then they had begun their series of documentaries on the Hopi Indians and on reservation poverty in California, which were more in line with KPFK's standard programming. The first of these combined recordings Ossman had made in 1965 for an unrealized documentary, readings from Frank Waters's *Book of the Hopi*, and materials provided by Craig Carpenter, a self-identified Mohawk messenger for the Hopi Traditionalists. An activist for the League of North American Indians (LONAI), Carpenter had connected with *Radio Free Oz* earlier in November, after hearing of an on-air conversation between Bergman and Ossman about the linguistic anthropology of Benjamin Lee Whorf and *Book of the Hopi* (the latter soon to become a touchstone of New Age spirituality). After helping to curate the first Hopi documentary for KPFK, Carpenter arranged for Ossman, Bergman, and Austin to be invited to attend the Soyal ceremony at Hotvella village on the Arizona Hopi reservation in late December. They stayed for several days observing rituals and conducting interviews.[35]

Returning from Hotvella and Third Mesa, all four Firesigns rang in the new year on KPFK with a live performance of Barbara Garson's *MacBird!*, which had just been published in the December issue of *Ramparts*. A key signal of the New Left's turn against Lyndon Johnson, *MacBird!* used the murder-and-succession plot of *Macbeth* as the basis of a conspiratorial satire of the Kennedy assassination, LBJ's rise to power, and escalation of the Vietnam War. Garson's play would premiere two months later at the Village Gate in New York, followed shortly after by a Grove Press box set that packaged a cheap paperback edition of the script together with a two-LP original cast recording.[36]

Varied though they are, the happenings of December 1966—the tape experiments, on-air improvisations, Indian documentaries, and the countercultural *MacBird!*—can all be seen as sources for the album the Firesign Theatre would spend much of the following year recording: *Waiting for the Electrician or Someone Like Him*. But not even the LP release of *MacBird!* gives a clear indication of how the group would soon be signed by the world's biggest record label and begin work on their first album at Columbia Square. That, instead, had everything to do with the fact that by 1967, Los Angeles had become the world capital of the pop music industry. Firesign's path led not through Grove Press's Barney Rosset—US publisher of Beckett, Ionesco, Garson, and Burroughs—but through the producer, songwriter, and architect of the "California sound," Gary Usher.

LISTENING TO THE ELECTRICIAN

Side 1

Gary Usher had made his name in early 1960s hot-rod and surf music with groups like the Surfaris and the Hondells, and he had been an early collaborator of Brian Wilson's, with whom he cowrote the Beach Boys' "409" and "In My Room." By 1967, Usher had become, like Wilson, a leading figure in the California sound's expansion and diversification. He had met Phil Austin in 1966 through "Duckman," a one-off *Batman* parody pitched by Austin and two friends, which Usher gamely produced as a seven-inch single for Decca Records. Noticing Austin's skill as a writer and voice actor, Usher decided to keep in touch about future projects, the next of which would be a novelty project, *The Astrology Album*, featuring cameos from the Byrds' David Crosby and the British expat folk duo Chad and Jeremy.[37]

The year 1966 had been the final one that the seven-inch single was pop music's predominant format. By early 1968, when *Waiting for the Electrician* was released, the youth music field had been reorganized around the LP, and Usher, when he came to Columbia in late 1966, became increasingly focused on the production of albums. For years, the record industry had expected that teenagers would abandon pop music for more serious "adult" genres as they got older.[38] Instead, it was pop music that seemed to grow up. After 1965—the year of the Beatles' cosmopolitan *Rubber Soul* album, the chamber pop of *The Beach Boys Today!*, the invention of folk rock, Bob Dylan going electric—commercial pop music could for the first time be seen as art, soliciting a more serious and durable regard from the fellow kids. This in turn elicited the neologism *rock music*, which Ellen Willis would cannily describe as a periodizing term, while Christgau, whose famous "Consumer Guide" was notably a survey of albums, would muse that "maybe what distinguishes rock from rock and roll is that you write criticism about it."[39]

Rather than rebelliously rejecting mainstream procedures, rock music appropriated many of its institutions and attitudes, foremost among them the long-playing record album.[40] The long-playing record album had historically been the domain of "adult" genres: classical music, jazz, Sinatra, Broadway cast recordings, the kitschy experimentalism of high-fidelity exotica. These conditions also explain why, shortly after Paul McCartney could be heard talking up an Alfred Jarry play he'd heard on the BBC Third Programme (Martin Esslin's production of *Ubu Cocu*), it would be possible to record a comedy album with literary pretensions for the rock audience.[41]

Those pretensions were a part of the genealogy of the LP, for in addition to its adult musical genres, the field of albums had been shaped to a surprising degree by nonmusical spoken word genres, comedy and literature most of all. The November 20, 1961, issue of *Billboard*, virtually a special issue on the phenomenon, featured numerous articles on both literary and comedy albums and added a helpful chart for people in the industry, showing the "strong commercial appeal" of these records together with albums commemorating historic news events. The chart included, in no particular order, albums by the comedians Stan Freberg, Rusty Warren, Moms Mabley, and Dick Gregory; seven different Shakespeare recordings; Dylan Thomas's visionary radio play *Under Milk Wood*; selections from the Nuremberg trials; Alan Shepard's space flight; and the George Martin–produced novelty record *Peter Sellers and Sophia Loren*.[42]

Given this history, it will be less surprising that the innovative Gary Usher, having just produced the Byrds' eclectic *Younger Than Yesterday* album for Columbia, would feel comfortable signing a fledgling comedy act, ostensibly to record a one-off album commemorating *Radio Free Oz* or

Figure 4. Gary Grimshaw's poster for the Elysian Park Love-In, March 26, 1967. Gary Grimshaw Legacy Foundation/Laura Grimshaw.

the unexpectedly successful Elysian Park Love-In, which had been largely organized on Bergman's radio program (fig. 4).[43]

Viewed from the other end of the telescope, it will also explain why each side of the resulting album—*Waiting for the Electrician or Someone Like Him*—would begin by citing a comedy album released in 1961. Though the B side's citation is more subtle, most of Firesign's audience would have instantly recognized that, before its cheeky citations of Ginsberg's *Howl* and *Rubber Soul*, *Waiting for the Electrician* began with a brutal riff on the widely known Capitol LP *Stan Freberg Presents the United States of America*. Still quite active in late-1960s LA, Freberg had been Jack Benny's successor on CBS radio in the 1950s and had a discography dating back to a famous seven-inch parody of *Dragnet* ("Duckman" in fact owed it more than a small debt). He had gone on to augment these successes with a parallel career in advertising. Some of the notoriety, however, of *Stan Freberg Presents the United States of America* ("Vol. 1: The Early Years") was because it acknowledged, rather than ignored, the exploitation of Indigenous people as the central fact of The Early Years. Somewhat to his credit, in 1961, Freberg featured Native American characters throughout side 1 of the album. This was clearly on Firesign Theatre's minds as they composed a far more acerbic comedic history, informed by their experiences on Third Mesa.

Freberg's record parodies iconic historical events and figures, from Columbus's voyage to Washington and the Battle of Yorktown, in skits and show tunes performed with the support of twelve voice actors and the Billy May Orchestra. The plot is spiked with funny anachronisms, as when King Ferdinand dismisses Columbus as a long-haired beatnik dreamer ("You mean 'old round, round world'? You and your Bohemian friends!") and then decides to tag along ("My doctor told me I should go to Florida for the winter"). The comedy is deflating (it turns out that Columbus's real dream is to open America's first Italian restaurant) but, following the storyline of a grade-school primer, is genial and in no meaningful way revisionist, which should not be surprising. More surprising is the extensive use of Indian characters (albeit played by white actors), whose status as colonized people forms the basis of the record's contemporary satires of postwar liberal hypocrisy. In this logic, Freberg's Indians are ciphers for contemporary white ethnic minorities, as when the mayor of Jamestown is counseled to take an Indian running mate or (more outrageously) when the 1626 sale of Manhattan becomes a satire of midcentury white flight in voices that combine stage Indian and borscht-belt stereotypes: "Too many moons we live here, White Cloud. Time to unload this crummy island!" It's hard to call this instrumentalization overtly hostile—and in his memoir, Freberg describes at length many subsequent encounters with the album's Native fans—but the fact that the Indians are left behind on the second side is evidence of the album's prevailing patriotism.[44]

Waiting for the Electrician is from the very beginning far more unpatri-
otically revisionist. A dozen years before Howard Zinn's *People's History of
the United States* would begin with the devastating story of the Arawak
Indians, "full of wonder, . . . sw[imming] out to get a closer look at
[Columbus's] strange big boat," Firesign's "Temporarily Humboldt County"
is a darkly parodic history of Western colonial extraction, told from the
perspective of the Indians.[45] Extending like Zinn's history into the present,
it also insists that the abuses of colonialism are ongoing and not merely
historical: Spanish conquistadors are followed by Catholic missionaries
followed by the wagon train followed by the railroad, by government
agents, tourists, hippies, and Hollywood casting services.

This shift in perspective takes place not only at the level of "historiography" but also aesthetically at the level of sound design. The experience
of hearing Freberg's record is to imagine actors moving (in impressive
stereo) across a Broadway stage. Listeners to "Temporarily Humboldt
County," by contrast, are positioned aurally alongside two Indian characters played by Ossman and Austin. Their soft voices mixed loud, Ossman
and Austin's characters occupy the foreground of the drama's imaginary
space. They greet each historical wave of colonists—"Aren't you the True
White Brother who's supposed to come live with us in peace?"—and then
recede before the much more active sound world that describes the settlers, who proceed with general indifference to the Indian characters. "My
fellow settlers!," shouts the Wagon Boss, "we stand here at the edge of civilization, on the banks of the Mississippi River, lookin' West, at our destiny.
What may appear to the fainthearted as a limitless expanse of godforsaken
wilderness . . . is in reality a golden opportunity for humble, God-fearin'
people like ourselves, and our families, and our children, and the generations a'comin' to carve a new life out of the American Indian!" Ossman
and Austin's Indians are a kind of audience within the drama, mirroring
the album's listeners, who are positioned to hear from the Indians' perspective—a technique for cuing sympathy and political alignment that
they had discovered in late 1930s radio.[46]

Further encouraging this sympathetic association is the fact that the
Firesign Theatre do not use stagey accents for the Native characters, even as
their panoply of white aggressor buffoons—Spanish conquerors, Irish missionaries, Anglo yahoos, senators who sound like W. C. Fields—together

make up a clownishly exaggerated Bakhtinian heteroglossia. This is an exact inversion of the vocal scheme of *Stan Freberg Presents the United States of America*, where the founding fathers all sound exactly like Freberg (which is part of the joke) and the Indians all sound like John Ford extras. The aesthetic choice is not without its difficulties though. Ossman and Austin's characters are stoic observers of five hundred years of a highly kinetic, exploitative, recursive history. "Remember what the Great Spirit said," intones Ossman's Indian character at the beginning and the end, "'Follow the peaceful way.'" While the political point of this aesthetic choice is clear—the violence of colonialism is reiterative and ongoing—the portrayal does not avoid the romantic effect of positioning the Indians in a nostalgic, even exterior, notion of time, an "exotic, mystical alternative to Cold War America" in which Ossman and Austin perform literally what Philip J. Deloria studied critically in his 1998 book *Playing Indian* ("a supposedly authentic Indian reality always located someplace outside American society") and risk perpetuating a form of static victimized identity Gerald Vizenor has countered with his concept of Indigenous *survivance*.[47]

Nevertheless, much of the specific politics of "Temporarily Humboldt County"—critiques of federal boarding schools, mineral extraction, Christianity—were received from materials provided them by LONAI and the Hopi Traditionalists (both of which were, to a degree Firesign did not at first comprehend, political organizations engaged in disputes internal to the Hopi reservation). Moreover, as the anthropologist Brian D. Haley has revealed, Firesign had also been granted access to a number of Hopi prophecies, with the idea that they might further disseminate them, and it is this political-religious material that provided "Temporarily Humboldt County" its deeper impetus and structure. Derived from *Wutatakyawvi*, a petroglyph panel near Orayvi, the Hopi Life Plan prophecy was believed to depict a message from the Great Spirit, or *Maasaw*. In Craig Carpenter's interpretation, which the group recorded on-site in December 1966, Maasaw's instructions for a "simple, humble life" had created a paradise that was later spoiled by outsiders. Arriving with a cross and a false life plan, the interlopers brought with them "insanity, wickedness, and obliteration."[48] The Purification Day prophecy was provided by the Hopi Traditionalist spokesman Thomas Banyacya, who was also interviewed by Firesign at Third Mesa. Banyacya's synthetic account foretold the day when the "True White Brother" (*Pahaana*) would return from the East to initiate a violent cleansing that would decapitate world leaders and cleanse the world of evil. "Temporarily Humboldt County" (a motto Firesign borrowed from Carpenter) is a synthesis of these prophesies. Ossman and Austin's Indians look to each wave of settlers as the hoped-for return of Pahaana; each generation proves instead to be a further deviation from Maasaw's Life Plan (the fateful cross being in one instance that of the Catholic missionaries, in another a "sacred antenna" that provides "better reception").

Haley's research discloses one further context for the piece: by the time they began recording *Waiting for the Electrician* in May 1967, the Firesign Theatre had already played a substantial role in instigating the counterculture's romantic fascination with the Hopi Indian. Following the December trip to Hopiland, KPFK broadcast two more multihour documentaries on January 26 and March 12, the latter leading to the March 18 Day of Discussion & Dialogue with Traditional Indian Leaders at the First Unitarian Church. The programs mixed field recordings and interviews, intensely committed historical and political critique, reports from political actions in the Northwest (the voice of the activist comedian Dick Gregory, arrested at Puget Sound, is the first thing heard in the January broadcast), and pieces based on the spiritual material Carpenter had been authorized to share with the Firesigns and with Bergman in particular, who disclosed the Life Plan and Purification Day prophecies on March 12. While these led to "direct ties on the ground" between the Indians and the counterculture, they also elicited further distribution of Indian material through the ur-hippie newspapers *Oracle of Southern California* and *San Francisco Oracle*, each of which published images of *Wutatakyawvi* together with Carpenter's gloss and separate romanticizing pieces about Indians.[49] All of this affinity-making—encouraged by Carpenter, both *Oracle*s, the newly formed Committee for Traditional Land and Life, friends of Stewart Brand, and Timothy Leary's associate Baba Ram Dass— was to culminate at the Grand Canyon on the summer solstice, at a mooted Hopi-Hippie Be-In. It would, supposedly, follow the examples of the January 14 Human Be-In (a "gathering of the tribes") in San Francisco's Golden Gate Park and the *Radio Free Oz*–inspired March 26 Love-In in Elysian Park. By the end of April, thankfully, vociferous opposition from both Banyacya and the Traditionalists' opponents in the Hopi Tribal

Council, and strident dissent from Emmitt Grogan of the San Francisco Diggers street theater group had begun to scuttle the plan.[50]

"Temporarily Humboldt County," in addition to its objective critique of colonialism, can also be heard as a way of working out the consequences of the popularization of Hopi, and the Firesign Theatre's responsibility for it. This is suggested by the piece's recording dates in May and June 1967, as the Be-In plans were collapsing, with a final session in November. Whereas Austin and Ossman's interest in Indian affairs had been primarily motivated by reasons of social justice, Bergman was at first drawn to the Hopi Traditionalists because of his interest in spirituality and mysticism, something Carpenter encouraged. As KPFK's "Wizard of Oz," he was also in early 1967 the group's only celebrity. After the Love-In (where he spent much of the morning in a tipi), Bergman gave an extensive interview for the *Oracle of Southern California*, where he spoke at length about the Hopis' apocalyptic prophecies, alluded to the mooted Be-In at the Grand Canyon, and traded in some of the most notorious fantasies of the day. Asked if he felt "that the spirit of the American Indian is being manifested through a lot of the young people today?" Bergman answered: "That's precisely how I feel. Not only that, I'll take it a little farther. I think the souls of Indians are coming back. I think they're coming back as the young people.... It's mystical as hell.... Also another thing that the Indians are responsible for ... is peyote. But, the thing that's particularly interesting about it ... is that it was given to the Indians by the Great Spirit as a means of contacting the Great Spirit long ago."[51]

But compare this interview with one of the final twists of "Temporarily Humboldt County." Amid a busload of tourists, and the dismaying return of Soaring Eagle from Indian School ("Call me Eddie, I'm an American now!"), a hippie approaches and says: "Hey, man. Don't let him bring you down, now. There's a lot of young people in this country just like myself who really know where the Indian's at. And don't worry, soon we're all gonna be out here on the reservation, livin' like Indians and dressin' like Indians and doing all the simple, beautiful things that you Indians do.... Hey, got any peyote?"

This satirical rebuke of ideas Bergman had espoused in the March *Oracle* is given additional weight by the fact that Bergman speaks the lines on the album in a new attitude of self-mockery. *Electrician*'s acknowledgment that the counterculture would become yet one more exploiter—*Radio Free Oz* having unthinkingly paved the way for the hippie invasion already under way at Third Mesa—is a form of reflexive self-criticism that could be found elsewhere (though by no means everywhere) in 1967. In May, Robert Jasper Grootveld (Bergman's associate in Amsterdam) disbanded the international Provo movement to foreclose the possibility of attenuation or selling out. On October 6, the Diggers staged a funeral in Buena Vista Park for "Hippie, devoted son of Mass Media."

Unlike the Diggers and Provo, the Firesign Theatre would continue on from within the belly of the beast of mass media, which they made their principal subject (though not to the point of recognizing modern media technologies' historical relation to the ethnographic study of Indians).[52] Nor did they openly disavow the Traditionalists, some of whom had vigorously opposed the Grand Canyon Be-In, even as they rethought their own representational strategies. Though they would not hesitate to put on other accents, Firesign would never again speak on their albums for or as Indians. In a caustic piece they performed at the Magic Mushroom club in November ("A Shadow Moves upon a Land," simulcast on *Radio Free Oz*), Firesign prefaced a withering self-critique with a multivoiced documentary poem in the style of Charles Reznikoff's *Testimony*, harrowingly reciting early nineteenth-century accounts of malaria epidemics and the ensuing mass deaths of Indians. Here, they inhabited the role of the historically implicated white people who were witnesses and recorders of the genocide:

DAVID: Since 1829 an intermittent fever has carried off vast numbers of Indians. . . . I have found the Indian population in the lower country below the falls of the Columbia, far less than I had expected. Since the year 1829.

PHIL P: 1829.

DAVID: 1829, probably seven-eighths have been swept away by disease.

PETER: So many and so sudden were the deaths which occurred that the shores were strewn with the unburied dead.[53]

They would perform variations of this distressing piece—intended to unsettle audiences expecting comic relief—at least through their East Coast tour in spring 1970.

As this rather long account indicates, "Temporarily Humboldt County" adapts and works with multiple sources in a way that can only be called literary, though this literariness could itself be subject to critique. (Galvanizing what came to be known as the Native American Renaissance, Kiowa writer N. Scott Momaday's *House Made of Dawn* would be awarded the Pulitzer Prize for Fiction in 1969.) But it is important to note that the nine-minute piece is only the first of three that make up the first side of the album. After the travesty of Freberg—the "historical" basis for what follows—the remaining tracks concern, respectively, the Summer of Love present and a speculative future that emanates forebodingly from the imagined coming institutional hegemony of hippiedom. "Ever since the sun took LSD, it's been a fundamentally better sun," proselytizes Tiny Doctor Tim (Leary) at the "Lazy O Magic Circle Dude Ranch and Collective Love Farm." This goofy solipsism finds its political consummation in the pothead authoritarianism that closes out the side, where kids who are not "hanging out in the parking lot relating with the rest of the kids and teachers" are "taken away for re-grooving" and the unhip developing world is bombarded by exploding copies of *Naked Lunch* (orders from "the Paisley House on Capitol Hill, man"). Taken together, Ossman would remember, "It has to be the bleakest comic portrait of America since *Huckleberry Finn*."[54]

Side 5

As deserving of exegesis as this may be, it is the second side of *Waiting for the Electrician* that is regarded as the more remarkable breakthrough—a side-length work that forms a clear template for the group's later albums. It can, though, be discussed more rapidly. Like the A side, it begins by citing a comedy record from 1961: Del Close and John Brent's *How to Speak Hip*. Following in the spirit of Close's *"Do It Yourself" Psychoanalysis Kit* (1959), *How to Speak Hip* was a send-up of the contemporary phenomenon of instructional records—albums like *Think and Grow Rich* (1960), *How to Handle Your Own Boat* (1962), and *Let's Hula* (1962), as well as Berlitz's language instruction records, which debuted in 1960. On *How to Speak Hip*, Close gives parodically authoritative instruction in beat culture argot, which Brent punctuates with parodically authentic *bon mots*: "Hello there, and welcome to the exciting world of hip. *Just relax baby me and this other cat we're going to straighten you out.* This is a new departure in language instruction for English-speaking people who want to talk to, and be understood by, jazz musicians, hipsters, beatniks, juvenile delinquents and the criminal fringe."[55]

The long piece "Waiting for the Electrician or Someone Like Him" similarly begins as a pastiche of a language instruction record. But whereas Close and Brent remain faithful throughout to the original fiction (it is always an instructional album), Firesign uses the conceit to generate a world beyond it:

ANNOUNCER: This is Side Five. Follow in your book and repeat after me as we learn three new words in Turkish:

"Towel"

"Bath"

"Border"

May I see your passport, please?

(Sound: the background of a large terminal.)

P: Yes, I have it right here.

GUARD: Look at this. This photograph doesn't look a bit like you now, does it, sir?

The vocabulary word elicits a "border" of a vaguely Eastern bloc country, the customs station of an airport or train station, customs officials, and a protagonist (technically unnamed but indicated in scripts as "P" after Kafka's "K"). Hassled about his passport, P goes to the information desk, attempts to send a telegram (which he is told will be sent to him in an hour), tries and fails to get a taxi, and is put up in a hotel, where, exiting an elevator, he is mistaken as a guest at a state dinner, whose honoree suddenly dies. He is then scuttled away by revolutionaries to "The Ice Show" at the Winter Palace, which for obscure reasons becomes consumed by violence (possibly an allegory of 1917); he then returns to the banquet, where he is heard surprisingly eulogizing the just-having-died Lord Kitchener in a bid to seize power for himself, whereupon he is arrested and sent to prison.

To the extent that it aims to produce a comically excruciating paranoia, "Waiting for the Electrician or Someone Like Him" is classic dope humor.

Its absurdist staging of bureaucracy, though—sign here, please, and here and here and here and here and here and here—is both Kafkaesque generally and a specific nod to Vaclav Havel's play *The Memorandum* (which had been translated and adapted for the BBC by Esslin in December 1966).[56] "Electrician" also evokes the narrative circularity that was a hallmark of the theater of the absurd. As the side ends, P exits a taxi and breathlessly sprints to the border, where he is again greeted by the voice of the customs agent/announcer: "Over here. This way. That's it. You've made it. Welcome to Side Six. Follow in your book and repeat after me as we learn our next three words in Turkish: Coffee, Delight, Border, May I seeee [tape slows] yooouuurrr ... passspooorrtt ... pleeeaaasssssee ...?" (Allegorically, we are still waiting for the electrician, but maybe someone's just unplugged the turntable.)

The best-remembered part of "Electrician" is the non sequitur event that surreally links P's imprisonment and his later taxi ride to the border. Having apparently passed out in a fit of hysteria in his cell, P is then revived and wheeled "into the Isolation Ward ... ready to play Symptom Six of *Beat the Reaper*!"—a leap that is exactly as jarring as P. T. Barnum's sudden appearance at the end of Michel de Ghelderode's *Christopher Columbus* (as performed on KPFK in 1966). Although to this point we appear to have been in an unnamed Eastern bloc country, and certainly somewhere in Europe, the drama is now suddenly fully in the realm of contemporary American media culture, on the set of a television game show. Each conscripted contestant on *Beat the Reaper* is injected with a virus and has to name the disease in order to be given the antidote and go on to the next round. P successfully proceeds past Symptom Six (jaundice) but is then unable to identify the Big Disease, which turns out to be the plague.

This is a reference to Artaud and to Bergman's associates in the Living Theatre (see my Liner Notes), and it may also refer to the 1947 novel *The Plague*, which Albert Camus wrote as an allegory of life under Nazi occupation in France and Algeria. If so, it is an augury of the Firesign Theatre's concern about the possibility of fascism in the United States, an idea we will see them exploring on their subsequent albums. While recording the album during the Summer of Love, Proctor and Ossman interviewed the artist, collector, and promoter Kate Steinitz for *Radio Free Oz*. As the

"Mama of Dada," the seventy-seven-year-old Steinitz had been at the center of the avant-garde arts scene in post–World War I Hanover with her collaborator Kurt Schwitters; with Schwitters escaping to Norway in 1937, Steinitz fled Nazi Germany for New York, settling in Los Angeles eight years later. Before discussing approvingly the recent Vietnam War protests, Steinitz was asked to describe her memories of the Nazis' escalating suppression and surveillance of artists and the German youth's embrace of Hitler. "Why did the young people fall into this so easily?" asks Proctor. "Because they got more than they had at home," answers Steinitz. "They got beautiful uniforms, they got music, they got all the entertainment they had in camp." They agreed there was something fascist in the LA police.[57]

If Camus uses the plague to dramatize the gradual acceptance of life under fascism, to make Camus's metaphor the vehicle of an entertaining game show is to stage as comedy the famous argument of two other LA exiles of Nazi Germany, Max Horkheimer and Theodor Adorno, whose own 1947 book warned about the authoritarian tendencies of what they dubbed the American culture industry. At the end of the record, as P speeds toward the border, we hear the entire society hysterically trying to catch the disease P was given on *Beat the Reaper*: "The plague's got the streets all tied up," says the driver, as the crowd bangs on the taxi's windows, "I wanna die! Touch me, touch me!" And on the radio: "Ed, it's an amazing scene here. Like lemmings the crowds are waiting on the shore, torches blazing, as the long line of shrouded funeral rafts drift lazily into view, great black candles flickering at helm and stern. The excitement is contagious, and so are the Black Cross volunteers as they pass from family to family, pausing now and again to touch a child's head." With its remediation of Camus, we might say, *Beat the Reaper* anticipates (in the spirit of Adorno) our contemporary culture of "virality."

In his 2013 study of the Grove Press, Loren Glass concludes his chapter on "Publishing Off Broadway" by discussing two plays Grove issued in cheap paperback editions in 1967: Tom Stoppard's *Rosencrantz and Guildenstern are Dead* (drafted alongside Bergman at Berlin's Literarisches Colloquium in 1964) and *MacBird!* (performed by the future Firesign Theatre and others on KPFK, New Year's Eve 1966). Though both were postmodern deformations of Shakespeare, Glass sees the plays as marking a "symptomatic division" at the heart of Grove's project of avant-garde popularization.[58] While Stoppard's play was promoted for classroom study, a strategy Grove first used for *Waiting for Godot* in 1954, *MacBird!* belonged to the current events activism associated with Grove titles such as Jack Gelber's *The Connection* (1959), the first major success for the Living Theatre. By 1967, these seemed irreconcilable tendencies.

If the Firesign Theatre can be seen as a third way, a successful mediation of avant-gardist and activist traditions, it has everything to do with their relationship to the medium and culture of the record album, which itself was suddenly at the center of youth culture after having been more closely associated with official culture and (as I will argue further) related to the book. *Godot* had itself been released as a double-LP cast recording on Columbia Masterworks in 1956, but by 1967 Grove could still only conceive of the LP as ancillary to a play's dramatic performance (as with *MacBird!*) and subordinate also to the printed text. They therefore missed the opportunity to recruit a new audience in tune with what the philosopher Bernard Gendron has called "popular music's turn toward an activist and highly appropriative stance toward high culture," its apotheosis having been the 1967 release of *Sgt. Pepper's Lonely Hearts Club Band* (an idea that occurred to McCartney within a year of his hearing *Ubu Cocu* on the BBC).[59]

To make the long-playing record album a dramatic work's raison d'être and condition of possibility, moreover—as the Firesign Theatre did—was in fact consistent with the longer tradition of "alliance[s] between modernist high culture and popular song" that Gendron studies in *Between Montmartre and the Mudd Club*, alliances that went all the way back to the emergence of Dada at Zurich's Cabaret Voltaire during the First World War.[60] More immediately, of course, this vinyl-centered literary practice emerged in alignment with changes in the field of pop music recording. We will finally return to this critical context now.

THERE'S A FOG UPON LA

What, exactly, did *Waiting for the Electrician* sound like? The fact that we can ask this question independent from the "literary" aspects of the

recording suggests that the relationship between them is not very adventurous. *Electrician* might be said to have a relatively straightforward "realism," meaning there is at every moment a foreground (voices) and a subordinate background (effects, ambient sound, music); no second hearing
is required to grasp the important jokes or plot details. Stereo separation
is panned hard left and right, as it was on many of the year's pop releases
(mono still being considered the primary format). Subsequent Firesign
albums, by contrast, will often feature several voices speaking at once (like
the simultaneous poetry readings at the Cabaret Voltaire or like talking on
the phone with the TV on), an effect abetted by an ever-increasing number
of recording tracks that would in turn allow for an increasingly complex
distribution of sounds mixed in fictional "space." Nonverbal sounds may
obscure or compete with speech, but each of these (and the competition
itself) will contribute to the semantics of the record. It is obvious that any
single hearing will be incomplete, but one is expected to follow at least one
of the strands.

Rolling Stone would remark on the second Firesign album's "quantum
leap in quality," and given the astonishing sound world of that album, it is
easy to overlook the fine details of *Electrician* (recorded, like *Sgt. Pepper*,
on four tracks).[61] Voices are sent through EQ filters to simulate forms of
mediation (telephones, a PA system, a cockpit radio); ambient background
notes cue the fictive locations (the hum of an elevator, the wind on the
desert, piano at a reception). There is always some kind of keynote (albeit
tactfully pulled back in the mix) and though these typically contribute to a
sense of realism, a few provide subjective, antirealist, or psychedelic
effects: experiments with tape speed, a disconcerting and increasingly
loud heartbeat, a horse that screams like an elephant, a backmasked trum-
pet reveille depicting a hit off a joint, a triangle wave tone of a very early
Moog synthesizer depicting the ominous operation of an electric chair.

It is likely that the Firesign Theatre originated many of these production ideas themselves (Proctor took a particular interest in the production
of sound effects, or "Foley"). But there can be little doubt of the importance of Gary Usher, who had gone from his early 1960s specialty of surf
and hot-rod music to recording and combining myriad genres as on the
"shining folk-rock, jazz-influenced pop, novelty space rock, and colorful
psychedelia" of the Byrds' transitional album *Younger Than Yesterday*,

released in January 1967.[62] Usher had also begun to experiment with theatrical production effects, notably on Roger McGuinn's paean to a quasar burst believed to be alien communication, "C.T.A.-102," the third song on *Younger Than Yesterday*. "C.T.A.-102" begins with two short verses addressed hopefully to the aliens—"Signals tell us that you're there / We can hear them loud and clear"—followed by a lengthy twenty-bar solo performed on an electronic oscillator triggered by a telegraph key.[63] After one more verse, the music abruptly stops before fading back in through a high-pass filter, as if to dramatize the song's outer space reception by the aliens, who comment approvingly in tape-speeded interstellar gibberish.

Usher had come to Columbia from Decca in November 1966, where he was for several months the label's only West Coast staff producer. Usher enjoyed unusual freedom to sign and record acts outside the supervision of the traditionally conservative label—first the Peanut Butter Conspiracy, then the Firesign Theatre—while updating the sound of acts like the Byrds. The Byrds, for their part, were drawn to Usher for his expertise with multitrack recording techniques (overdubbing, phasing, flanging, backmasking), which distinguished the sound of *Younger Than Yesterday* and its follow-up *The Notorious Byrd Brothers*, which began tracking in June. Usher, for his part, appears to have found inspiration in the weird conclusion of "C.T.A.-102," as many of his productions in the coming months featured similar unconventional experiments with song form. These culminated in the Usher-fronted studio group Sagittarius, whose first single, "My World Fell Down," devolves after two minutes of eerie sunshine pop into a sound collage of a crying baby, horse race, alarm clock, and bull fight (and nothing else) before a low-key a cappella section leads to a final reprise of the chorus. He repeated the trick on the second Sagittarius release, "Hotel Indiscreet," a saccharine ditty about a low-rate sex hotel into which Usher interpolated, in place of a conventional bridge, the voice of a radio preacher that morphs into a military drill officer.[64]

The interpolated voices on "Hotel Indiscreet" were provided by Peter Bergman and the Firesign Theatre, which was just one of many times that Usher called on the group to produce theatrical textures for his psychedelic productions. They made a tape collage of gunfire and explosions, unsettling the pastoral mood of David Crosby's protest song "Draft Morning" on *The Notorious Byrd Brothers*; they added parodic media voices to Chad

Figure 5. The Firesign Theatre at CBS Columbia Square Studio, 1967. Firesign
Theatre Collection, National Audio-Visual Conservation Center, Library of Congress.

and Jeremy's sanctimonious "Progress Suite" on the concept album *Of
Cabbages and Kings*; and they did further work for Sagittarius. Such com-
missions, since Usher's original pitch to the group seems to have been for
a novelty record in the vein of *The Astrology Album*, may seem to point to
a general uncertainty about the Firesign Theatre's identity (or autonomy).
But it also points to the Firesign Theatre's proximity to LA's pop music
scene, since Usher was in fact using the group as a kind of paramusical
version of the Wrecking Crew, the legendary LA session musicians who
had played on scores of the decade's pop recordings, including those of
apparently autonomous groups like the Byrds and the Beach Boys (fig. 5).

This, in turn, underscores another distinguishing feature of *Waiting for
the Electrician*'s sound, which is that its musical cues are performed by
these same session pros (including Glen Campbell), who grace the pro-
ceedings with a professionalism that is impressive though in retrospect
sounds rather conventional. Subsequent Firesign albums would have a
much more variegated sound world, composed of licensed prerecorded
music called "needle drops," obscure found recordings, self-produced

cut-ups, and performances from a few sympathetic instrumentalists in Firesign's inner circle. Though the Wrecking Crew gamely execute an out-of-tune rendition of "Hail to the Chief," their dissonance sounds mannered compared to the gleeful incompetence of the amateur Portsmouth Sinfonia's *Also Sprach Zarathustra*, which would close out the first side of Firesign's *Everything You Know Is Wrong* in 1974.

Maybe the most obvious thing to say about *Electrician*'s sound, which would be true of all their albums, is that it featured an array of voices that were made by four (and only four) people, realizing what Bergman would recall had been his inspiration, surrounded by massive billboards for *A Hard Day's Night*, at Piccadilly Circus in 1964: a four-man comedy group that was like a band. That model was given further inspiration in the form of the early mixes for *Sgt. Pepper*, which were circulating unofficially throughout LA in the spring of 1967, including at KRLA-AM, which had become the new home to *Radio Free Oz* after the Easter Sunday Love-In.[65]

Ossman has described how the group was inspired to make a "comedy record that you would listen to as many times as we were listening to *Sgt. Pepper . . . explicitly*," borrowing liberally from its production palette (as well as, maybe, gesturing to the famous acronym of "Lucy in the Sky with Diamonds" in their own "Temporarily Humboldt County").[66] We will explore Firesign's identification with the Beatles thoroughly in the next chapter, but it is worth mentioning here a major circumstantial connection: both groups enjoyed unlimited studio time to make their albums. For the Beatles, this had begun with the extensive overdubs and reduction mixes used for *Revolver* in 1966 and was licensed by the group's enormous success. In Firesign's case, the infinite studio time was granted by Columbia as compensation for the fact that, as a nonmusical act, they were unable to recoup publishing royalties called mechanicals.[67]

The new studio arrangements were perfectly, almost causally, suited to a world in which the album was becoming the aesthetic and economic center of popular music. *Sgt. Pepper* cued the attentive listening practices appropriate to the new aesthetic by eliminating spaces between tracks, printing lyrics on the back cover (presumably for reading while listening), and declining to issue a single from the album. As music historian Elijah Wald puts it, Beatles records were now to be understood as "fully conceived, finished objects in the same way that a book or a painting is a fully conceived, finished object"; they were "musical novels, designed for individual contemplation in their entirety."[68] Here, at last, is the short answer to the question we began with (does molding a comedy album in the form of the phonograph make it literary?), though Wald, writing in a book titled *How the Beatles Destroyed Rock 'n' Roll*, views it as a negative development. In his view, it had been in live performance that rock 'n' roll realized its most promising form of social and musical integration, "in which white and black musicians had evolved by adopting and adapting one another's styles" (12, 239). After 1966, in the place of this promising form of sociability and exchange, "bands can play what they like in the privacy of the studio, and we can choose which to listen to in the privacy of our clubs, our homes, . . . our heads" (246–47).

Wald extends a long tradition of antirockist Beatle apostasy, and there is a version of it in sound and media studies as well, in which the technological sublime of stereo is subject to the critique Wald gives recording studios and musicians. In many ways, the key term connecting them is *privacy*: private space, private experience, and the private possession of culture. Creative of an "orchestrated, technologized, managed sonic world," Jonathan Sterne has characterized high-fidelity audio as a scene of "situated omniscience, one that extends a particular kind of Enlightenment subject into the listening space . . . stretch[ing] toward a subjectivity that is at once somewhere and everywhere."[69] With origins in nonmusical devices like the binaural stethoscope, such listening has a history that includes the suburban living room of 1950s hi-fi ads, Maxell cassette tapes' legendary "Blown Away" campaign of the 1980s, and the present-day affect management associated with noise-cancelling headphones.

Objections to the categorical nature of the Sternean argument could be found in Alexander Weheliye's *Phonographies*, which, though similarly critical of suburban white normativization, affirms the desires of the heroine of the 1994 film *I Like It Like That*, "one of [whose] main goals is *to create a private sonic space for herself*" by listening to her Walkman or blasting the radio in the bathroom.[70] And an exception to Wald's broad judgment can be found in Ben Piekut's biography of the experimental rock group Henry Cow, children of the album era for whom "the effects of vinyl went far beyond multistylism to foster a radical cosmopolitanism and . . . experiments in group authorship that amount to more than the production of cultural commodities and the maintenance of ideological control."[71]

My point is not to dispute the antiphonographic arguments on their own terms but rather to notice how nearly the terms resemble ones from another, seemingly unrelated, phase of media studies—namely, the book history and history of reading of the late 1980s and 1990s. In the seminal 1989 essay "The Practical Impact of Writing," the book historian Roger Chartier studies the post-Gutenberg "spread of literacy, the widespread circulation of written materials . . . and the increasingly common practice of silent reading" that fostered an "altered relation to the written word [that] helped to create a new private sphere into which the individual could retreat, seeking refuge from the community." This established "new models of behavior . . . that tell us a great deal about the process of privatization in the early modern period."[72] "Personal communion with a read or written text liberated the individual from the old mediators, freed him or her from the control of the group, and made it possible to cultivate an inner life" (121). The library or study eventually became "a place where one can see [i.e., through congress with books] without being seen" (130).

With its invention of modern notions of privacy and an "inner life" distinct from the mediators of the church and the state, the personal library gave birth to the Enlightenment subject Sterne will find sitting in front of his stereo system, practicing the kind of immersive listening that would be made a vocation by someone like Christgau. From this perspective, it is fitting that one of the most interesting essays to follow the path of Sternean critique, Gustavus Stadler's "'My Wife,'" is about Andy Warhol's ardent use of his portable Norelco audiocassette recorder in the mid-1960s, work that notably resulted *in the production of an unreadable book*—four hundred pages of amphetamine-fueled conversations with the Factory personality Ondine, transcribed at the Factory and published in 1968 (by Grove Press) under the title *a: a novel*. In Stadler's reading, the "muddled, imprecise, oddly formatted, reader-unfriendly transcription" of the noisy tape manifests Warhol's playful critique of the cultures of immersive, isolated listening associated with album culture and multitrack recording.[73] It is just as obviously an experimental challenge to the cultures of literacy, reading, and the book studied by Chartier.

THE DIALOGIC IMAGINATION (SLIGHT RETURN)

In a slightly later essay, "Labourers and Voyagers," Chartier stresses that "attention should particularly be paid to ways of reading that have been obliterated in our contemporary world: for example reading out loud in its double function—communicating . . . and binding together the interconnected forms of sociability which are all figures of the private sphere."[74] In this spirit—which brings to mind the social and collective listening cited above and associated with the Firesign Theatre but obscured by the Maxell advertisement and wholly absent from the sound studies critiques—we can rescue a bit of truth from the degraded world of advertising.

Record companies, independently from such phenomenological arguments about listening and reading, looked to the practices of book publishers as they devised "long-playing" strategies to help manage the unpredictable sales patterns of albums, which were much more erratic compared to those of singles (whose temporality was more like magazines).[75] Not only might the value of a particular release be at first uncertain, as would famously be the case for the Rolling Stones' *Exile on Main Street* in 1972; it might also require reference to past work, a process the label could support by maintaining an artist's back catalog, in the hopes that customers would be slowly building their own record "libraries."

Though the model of the book was a matter of reference internal to record companies, Columbia made the analogy explicit with two advertisements in 1970. One of them marketed the Firesign Theatre's dystopian third album, *Don't Crush That Dwarf, Hand Me the Pliers*, by alluding to George Orwell's *1984* ("This album is 14 years ahead of its time"), the other simply called the epochal album *Bitches Brew* "A NOVEL BY MILES DAVIS" (its photo of the opened gatefold sleeve evoking a book).[76] Particularly because these artists will wind up together in the 1992 issue of *Motorbooty*, it is worth considering what we might learn by taking this confluence as something more than the cheesy bullshit of the same PR department that had authored the canonically risible "America listens while the establishment burns" and "the man can't bust our music" campaigns in the late 1960s.

Firesign's Columbia period and Miles Davis's electric period (also on Columbia) are for starters not roughly but *exactly* contemporary. In his landmark essays for *Downbeat* in 1983, Greg Tate traced the origin of Davis's still divisive "electric period" to the sessions that produced the acoustic *Miles Smiles* LP in October 1966, a month or so before Proctor reconnected with Peter Bergman and met Phil Austin and David Ossman at KPFK.[77] At the end of 1967, as the Firesign Theatre was finishing the mix of *Waiting for the Electrician* at Columbia Square in Hollywood, Davis was recording "Circle in the Round" at Columbia's 30th Street Studio in New York, his first track to exhibit, in the words of music writer Paul Tingen, "the musical influences of the '60s counterculture, his search for [a] dense and complex bottom end, and the application of postproduction technology."[78] Further committing themselves to multitrack recording, Miles Davis and the Firesign Theatre each released pathbreaking "electric" records in 1969 (*In a Silent Way* and *How Can You Be in Two Places at Once When You're Not Anywhere at All*) and did it again with the career-defining albums of 1970. Both acts sold out Carnegie Hall in the spring of 1974, released uncompromising climactic albums in 1975, and then entered years of recording silence—Davis by choice, Firesign by fiat.

Both Firesign Theatre and Miles Davis engaged with rock music in idiosyncratic, individuated ways that distinguished them from others who were motivated to do the same thing, though were often pandering or gimmicky (*vide* Marshall McLuhan's *The Medium Is the Massage* [1967] and J. Marks and Shipen Lebzelter's *Rock and Other Four Letter Words* [1968]). For all their obvious differences, Firesign and Miles Davis both responded to the hegemony of the psychedelic album era with recordings that combined group improvisation with postproduction techniques like the ones Gary Usher brought to his Columbia groups ("overdubbing, varying tape speed, playing the tape backwards, cutting up the tape, adding effects, and so on").[79] In Davis's catalog, *Bitches Brew* is a clear watershed for these practices, an album now recognized for its unprecedented confluence of musical *voices*: "jazz and rock, classical and African, improvised and notated music, live playing and postproduction editing."[80]

In 1991, the musicologist Gary Tomlinson read this achievement under the banner of Bakhtinian dialogism and heteroglossia. Remarking on "the rich dialogue of musical voices that went into the making of Davis's new styles around 1970—voices . . . from outside as well as inside the walls of 'pure' jazz," Tomlinson stressed that this sociality was produced as much through performance as through the postproduction techniques performed by Davis and his producer, Teo Macero:

> The dialogics of fusion involved also an extraordinary freedom of colloquy within the ensemble, a freedom of exchange wherein the complex bottom and any of its constituents could function as top and vice versa. This far-reaching democratization of the ensemble has an obvious antecedent in various versions of free jazz; but it came to Davis's music also from pop sources—specifically Sly and the Family Stone, whose vocal and instrumental arrangements featured all the musicians with an equality rarely heard in earlier rhythm-and-blues styles. And the democratic, multiple improvisations of much of Davis's fusion music took encouragement also from European avant-garde trends—specifically the structural indeterminacy sometimes explored by Karlheinz Stockhausen, whose music and thought Davis came to know around 1970.[81]

Much of this précis (complex bottom notwithstanding) could define the way the Firesign Theatre worked as well, especially the idea of the democratization of the ensemble. As a rule for their collective writing, Firesign agreed that every decision required consensus. Though this surely was not the case for mixing decisions on Miles Davis's records, it does point to the way both performance and multitracked studio recording realized a form of heteroglossia that conforms with Bakhtin's theory—"the novel as a whole is a phenomenon multiform in style and variform in speech and voice" (Bakhtin, 261)—thereby lending unexpected authority to Columbia's advertisement.

Just as the self-evident literariness of the Firesign Theatre made their records much more explicit realizations of the book archaeologically embedded in the concept of the album, their work was also much closer to the root of Bakhtin's magisterial genealogy of the novel. Bakhtin, we recall, famously extended his "prehistory of novelistic discourse" back to the "parodic-travestying" social practices of the medieval carnival and early modern commedia dell'arte, the latter defined as the "comedy of dialects" whose "laughter and polyglossia had paved the way for the novelistic discourse of modern times."[82] In Firesign's case, this took the form of the Renaissance Pleasure Faire and May Market, which originated in Los

Angeles in 1963 and became one of the touchstones of the Southern
California counterculture (the first issues of the *Los Angeles Free Press*
were produced as broadsides for the Faire). It was a kind of participatory
theater that, especially in its early iterations, thrived on both verisimili-
tude and anachronism, a spirit that film historian Thomas Elsaesser
would also later find to animate the creative explorations of media archaeology: "either despite, or because of its supposed obsolescence, [it became]
the repository for that different kind of future that seems to lie at the heart
of media archaeology's utopian aspirations."[83]

This connection to media archaeology would seem fanciful were it not
for the fact that the Renaissance Pleasure Faire had its very origins in the
culture industry, as Rachel Lee Rubin has shown. The Faire was originally
devised by Ron and (especially) Phyllis Patterson. Phyllis had taught
theater to children for several years at the Wonderland Youth Center in
Laurel Canyon, where she found that "one of the most popular theatrical
forms with the children was commedia dell'arte."[84] But Phyllis remembered that as it became institutionalized, "the Renaissance faire was able
to flourish thanks to the Hollywood blacklist, because the blacklist had the
effect of making gifted and skilled people [set builders, costume makers,
musicians, screenwriters], many of whom were left-wing activists, available to lend their talents first to the backyard classes and then to creating
the full-scale faire—thereby fostering an understanding of the connections between politics and culture for a next generation."[85] The Faire thus
drew on the expertise of Hollywood culture workers who had deep and
specific ties to the Old Left.

Neighbor to the Pattersons on Lookout Mountain Avenue, and with
children who attended the Wonderland school, David Ossman hosted a
five-hour broadcast from the Faire on KPFK (in costume), was anointed
Herald of the Faire in 1964, and participated every year through the
1960s. The Firesign Theatre performed at the Faire in 1968, shortly after
the release of *Waiting for the Electrician*.[86] It had by then penetrated
deeply into the Los Angeles counterculture. The Byrds' David Crosby, who
recorded songs about the Faire and about the March 1967 Love-In,
adopted the Renaissance look on the cover of 1965's *Turn! Turn! Turn!*,
and it can also be seen as a dominant fashion at the Love-In (fig. 6). Writing
in the *Freep*, John Wilcock's "Love-In Inventory" noted that it was "Very

Figure 6. Attendees of the Elysian Park Love-In, "clothes mostly in Renaissance Faire style," March 26, 1967. Photo by Jerry De Wilde. Permission of the artist.

Medieval in content with the banners, pennants, different encampments. Clothes mostly in Renaissance Faire style with some embellishments."[87]

Though the Byrds' manager, Jim Dickson, had helped to organize the Love-In, it is notable that contemporary accounts—such as Les Blank's documentary *God Respects Us When We Work, But He Loves Us When We Dance*—make only passing reference to the bands who performed (the Turtles, New Generation, Rainy Daze, and the Peanut Butter Conspiracy). In this way the Love-In differed from the event that had inspired it, San Francisco's Human Be-In, which featured performances by the Grateful Dead, Jefferson Airplane, Big Brother and the Holding Company, Blue Cheer, and Quicksilver Messenger Service, and is generally understood as an outgrowth of the rock-oriented multimedia events like Ken Kesey's acid tests and the Trips Festival of January 1966. In his book *The Democratic Surround*, media scholar Fred Turner sees the Human Be-In as both the culmination and the end point of a series of experiments in "imagining the relationship between media, polity, and self" that had their origins among antifascist American intellectuals in the 1930s.[88] By the psychedelic 1960s, these had become deeply informed by Marshall McLuhan's hopeful prediction that electronic media would supplant the "linear, hierarchical, psychological, and social arrangements that

McLuhan claimed print had brought into being" and bring about a phase of a neo-tribal "global village" instantiated by the total environment of electronic communication, "with sights and sounds that threatened [audiences'] ability to reason" (Turner, 285).

With its anachronistic origins in the "prehistory of novelistic discourse" and with its emphasis on the album—a medium that could be immersive in its articulation of sound yet inscriptive and iterative like a book—we can see the Firesign Theatre as a kind of inspired compromise formation to this problematic, a solution that is not a McLuhanite "secondary orality" precisely because it does not transcend the medium of writing but deeply instantiates it.[89] The ambiguity and promise of the album as a written form that could be theatrical and even hallucinatory, while also inspiring repeated collective listening and discussion, may also have been a way of retaining the critical potential Turner sees as beginning to wane as much of the psychedelic project turned to "raw mysticism" (284). Against the utopian determinism of McLuhan, as it turned more explicitly to its investigations of media on subsequent albums, the Firesign Theatre retained the possibility of investigating media as a form and source of authoritarian politics by maintaining a commitment to the vocation of writing.

2 RADIO / Duplicity Is the Double of Duality

How Can You Be in Two Places at Once When You're Not Anywhere at All (1969)

PUT-ONS AND PROPAGANDA (HOW TO BE IN TWO PLACES AT ONCE)

One Sunday evening in early 1970, Phil Proctor read a letter live on the Firesign Theatre's new program on KPPC-FM in Pasadena:

Personal:

Dear Mr. Proctor,

The President has asked me to thank you for the recording you recently sent him. He is pleased to have this humorous selection for his music library. Your generous expressions of friendship mean much to the President and he is glad to know that you and the cast members of the Fireside [*sic*] Theatre support his efforts for an honorable and lasting peace in Vietnam. He is grateful also for your kind personal sentiments.

   With the President's best wishes, sincerely yours,

Rose Mary Woods,
personal secretary to the
President.[1]

The letter was evidence of a successful put-on. The president (Richard Nixon) had campaigned on ending the Vietnam War and then escalated it once he was in office; the recording (the Firesign Theatre's second album) contained a corrosive allegory of the war and the culture that produced it. The KPPC audience knew the album and would have understood that Proctor had been soliciting just such a letter.

The put-on was the late 1960s' distinctive mode of performative satirical critique and one of the counterculture's hallmark forms. In August 1968, just before the Firesign Theatre began recording Nixon's "humorous selection," the Yippies nominated a pig for president at the Chicago Democratic National Convention (a year earlier Abbie Hoffman had promised to levitate the Pentagon). A few days later, feminists greeted the Miss America pageant in Atlantic City by crowning a sheep on the boardwalk. In 1967, Pauline Kael had famously written that the stylized violence of *Bonnie and Clyde* had made it "the first film demonstration that the put-on can be used for the purposes of art."[2] A year before that, the Firesign Theatre's first appearance had itself taken the form of a put-on. They pretended to be filmmakers implausibly screening their work live on the radio and then staged the film's censorship in real time. To their surprise the feigned suppression elicited dozens of outraged phone calls to KPFK, an experience David Ossman said showed them the power of radio, just as *The War of the Worlds* had done for Orson Welles.[3]

The put-on is a phenomenon of modernity and establishes its cultural meaning through modern technologies of transmission and mediation.[4] The most famous put-on of them all, Welles's *War of the Worlds*, was a radio fiction that played with the very new broadcast technique of breaking news bulletins.[5] The Firesign Theatre's Nixon/Vietnam put-on involved an LP album, FM radio, and the postal service. "By annihilating distance, the letter allowed the person who wrote it to be in two places at once."[6] Its communications circuit was completed not with the uncomprehending letter Woods wrote for Nixon but with Proctor's gleeful reading of it to the comprehending audience of the Firesign Theatre's *Radio Hour Hour* on KPPC.

Or, rather, so it appeared to have been completed. A second letter, written fifteen years later, reveals that the put-on's communications circuit is often much more volatile. That letter appeared in *Four-Alarm FIRESIGNal*, the unofficial Firesign Theatre fanzine founded by Elayne Riggs in the mid-1980s. From the start of the group's long hiatus in the mid-1980s until the establishment of the alt.comedy.firesign-thtre Usenet group in the 1990s, *"FAlaFal"* (sorry) was the primary means of communication for the Firesign fan community, a space for trading information about locating out-of-print recordings, just-discovered references (some of which were themselves put-ons), or tales of old listening sessions—especially "first time" stories. Chris Ward's December 1985 letter is characteristic; it marvels at the literary world-building of the Firesign records (quoting Ezra Pound, "It all coheres!") and expresses appreciation for the way the zine has connected him with other fans. It then concludes very atypically: "I'm working for the U.S. Information Agency now, and I'm constantly surprised by the number of colleagues here (and in our sister agency the State Department) who are lovers of Firesignia. . . . I'm being sent off to Costa Rica in six months, and you can bet your dead presidents I'll be bringing along some o' that gas music from Jupiter."[7]

What does Ward's letter—which was not a put-on—tell us about the put-on, or about the Firesign Theatre as practitioners, theorists, instruments of the put-on? Founded in 1953, the United States Information Agency (USIA) was the Cold War public relations (or propaganda) bureau tasked with "telling America's story to the world." It sponsored art exhibitions, libraries, and book programs; administered the Fulbright Scholarship program; and oversaw the Voice of America radio service. Targeted at its inception by Joseph McCarthy, its director during the Kennedy administration was the venerated World War II broadcaster Edward R. Murrow. During the Vietnam years, when the phrase "public diplomacy" became an official euphemism for propaganda, the USIA was a major arm of the anticommunist "hearts and minds" campaign, and it dramatically increased activity in Southeast Asia after LBJ committed ground troops in 1965. Together with the Joint United States Public Affairs Office (JUSPAO), the USIA coordinated psychological operations (PSYOPs) for the war, airdropping leaflets, blasting messages inland from loudspeakers on ships, and organizing radio programming.[8]

Chris Ward's fanzine message testifies to the abiding political importance of radio after Vietnam; assigned to Costa Rica in 1986, Ward would have been tasked with directing radio transmissions toward Managua, Nicaragua, in support of anti-Sandinista Contra rebels.[9] But his claim that the Firesign Theatre was popular among US diplomats and propagandists indicates something more—namely, the close proximity of the culture and techniques of propaganda to the culture and techniques of the put-on. Identifying the put-on's relation to disinformation had in fact been one of the official interpretations of 1938's *War of the Worlds* phenomenon; it was a mere matter of weeks before the director of the Princeton Radio Research Project, Hadley Cantril, secured funding to study how public belief in the Martian invasion might help evaluate American susceptibility to Nazi and other fascist propaganda.[10] In another key, the same insight explains the notorious claim of Rush Limbaugh—demagogic propagandist / master prankster—to being a Firesign Theatre fan.[11] Kael seems to have been aware of the same possibility when she wrote that "during the first part of [*Bonnie and Clyde*], a woman in my row was gleefully assuring her companions, 'It's a comedy. It's a comedy.' After a while, she didn't say anything."[12]

The instability of the put-on, the reversibility of its humor, was a hazard inherent to the form (think of the stories from *The Onion* that circulated globally as actual news, preceding and inspiring the onslaught of disinformation we now take for granted).[13] This was something the Firesign Theatre would come to understand better than most. A 1971 letter from a teenager to syndicated advice columnist Ellen Peck is further confirmation. On Peck's advice, he played for his father *Waiting for the Electrician*, "the piece about how the Indians were cheated out of their land. . . . Dad missed the point entirely. He thought it was the funniest thing he ever heard. Now he plays it for all his friends and tells them what good taste his son has in comedy."[14] Peck advised him to give him a copy of *Bury My Heart at Wounded Knee* and keep trying. In fact, a principal theme of the album the Firesign Theatre had sent to Nixon was this very idea—that the put-on was propaganda in a different context—and it is one way of understanding the psychedelic doublespeak of its title: *How Can You Be in Two Places at Once When You're Not Anywhere at All*. That discovery was one conclusion of the examination of radio that the group made as they recorded the album, a media-archaeological process that turned them inward to the meaning of their situation in the studio (and its history in the fight against fascism in the 1930s and 1940s), and outward to the multifarious uses of radio in and around the Vietnam War.

Then and now, Vietnam has been understood as the "first television war." In 1968, Marshall McLuhan wrote that "the television war has meant the end of the dichotomy between civilian and military [and] the main actions of the war are now being fought in the American home itself," while at the same time insinuating that the war had been a crash course in westernization for the Vietnamese.[15] (It is in this context that critic Sidney Finkelstein attacked McLuhan's method as a "form of pulling the reader's leg"—a bad-faith put-on masquerading as theory.)[16] The Firesign Theatre's second album, on the contrary, suggests that radio had remained the more profound way of "being in two places at once" because it illustrated uncomfortable continuities with the Second World War—specifically, the US's application of techniques learned from the Nazis—rather than a wishful progression beyond it. A recent study avers that Vietnam "saw the most intensive use of psychological warfare in history."[17] At the same time, however, an efflorescence of radio stations on the formerly moribund FM band would constitute a vital information network for the antiwar movement in the United States. Many of those stations, until around 1972, also regularly broadcast the Firesign Theatre.

That this dialectic would have been highly apparent to the Firesign Theatre can be represented by the career of Phil Austin, who came to the Pacifica station KPFK after receiving his radio training from the 360th Psychological Warfare Unit of the US Army.[18] As if to show the reversibility of these roles, before joining the US Information Agency, Chris Ward had been the host of the *Breakfast Special* on WORT, Pacifica's FM station in Madison, Wisconsin.[19]

LABEL TROUBLES

There had almost been no second Firesign Theatre album. Columbia Records released *Waiting for the Electrician or Someone Like Him* in January 1968, and many anecdotes suggest that the album found key audiences quickly. Stoned newsmen are said to have listened to it while monitoring the Tet Offensive from the roof of the US Embassy, days after Bob Fass played it all night on WBAI in New York.[20] The future KCRW DJ Roger Steffens remembers hearing it in early 1968 on his first tour of duty in Saigon.[21] Another serviceman, quoted in a *Rolling Stone* article on drugs in the military, said he found the album at the PX filed among the children's records (was *that* a put-on?).[22] For most of the year, though, such reception remained strictly underground. It wasn't reviewed in *Rolling Stone,* and *Creem* did not yet exist. It wasn't played on AM radio. Sales were very slow, and the label was unimpressed.

Columbia grew still more concerned when they heard the wildly experimental thirty-four-minute recording the Firesign Theatre had gone on to make that September, a piece that was much further out than anything on *Electrician* and much more grimly topical. A&R reps were played a rough mix late in the year and decided not to renew the group's contract. Finishing the album required the intercession of august figures from the label: the legendary talent scout John Hammond, who had worked with everyone from Billie Holiday to Bob Dylan, and the classical producer John McClure, who had overseen important recordings with Stravinsky, Leonard Bernstein, and Bruno Walter. In late 1968, Firesign's manager, James William Guercio, brought the outsider composer Moondog to McClure's Columbia imprint Masterworks.[23] In January, Masterworks sponsored an editing session that cut six minutes from Firesign's new recording, fitting it to a single LP side.[24] In February, another long piece was recorded for the flip, a pastiche of 1940s radio detective fiction titled "The Further Adventures of Nick Danger." Columbia released the album in July 1969. By October, it had made the *Billboard* charts and secured the group's medium-term future with the label.[25] By the end of the year, it had been reviewed together with *Electrician* in both *Rolling Stone* and *Creem.*

Recording had begun just after the mayhem at the Chicago DNC; it finished in the shadow of Nixon's inauguration. The title *How Can You Be in Two Places at Once When You're Not Anywhere at All* at first seemed like a blithely inane riff on Taoist duality (which it also was); with further listening, it became redolent of the duplicity, confusion, unease, and violence from the last ten months in Vietnam and the United States. It also, and relatedly, asked a question whose answer could have been media, propaganda, or simply *radio.*

CONTRADICTORY SPACE

Viewed one way, the record's success confirmed Hammond's and McClure's professional intuitions. The prestigious imprimatur of Masterworks encouraged the Firesign Theatre to make an album that not only sustained repeated listening but actively required it. Robert Christgau advised readers to listen to the second side first since "the title side is so far out it lacks credibility alone," but he also called it "close to a work of genius" and awarded it a very rare A+.[26] As Christgau's career itself attested, there was now an audience acculturated to the requisite immersive, associative, and concentrated listening ("quietly, stoned perhaps, in the company of a few friends, on a sound system that can convey its technological nuance").[27] Coextensive with the new importance of the LP format, this (apparently) new listening practice was part of what distinguished the culture of the neologism *rock* from the earlier, dance-oriented adolescent culture of "rock 'n' roll," a transition consecrated with the release of *Sgt. Pepper* in summer 1967.[28]

As Xgau notes, this culture was also enabled by technological and industrial changes, which in turn made possible the albums of the Firesign Theatre. These changes sanctioned and regularized techniques that had transformed the meaning of sound recording and were prioritized by the LP format (as discussed in chapter 1). Following on the innovations of high-fidelity recording, which allowed engineers and producers to construct "a soundscape specifically for the home listener," the 1960s expansion of multitrack recording had enabled the possibility of what the musician Brian Eno would call "in-studio composition."[29] Bouncing, overdubbing, and editing tracks allowed musicians to abandon the faithful simulation of a single time and space of performance, at first gradually (as in high-fidelity classical and exotica recordings of the 1950s) and then dramatically (most notably in psychedelia).[30] Hence the path that led the Beatles from the string quartet of "Yesterday" and sitar on "Norwegian Wood" to the ludic diegetic sound effects on "Yellow Submarine" to the esoteric sound worlds of "Strawberry Fields Forever" and finally outward to an album that pretended to feature a metafictional Lonely Hearts Club Band ("the most brilliant fake in rock and roll history," said Beatleologist Devin McKinney).[31]

Eno concluded that the studio now "put the composer in the identical position of the painter," and in their study of aural architecture, Barry

Blesser and Linda-Ruth Salter extend the simile. Emphasizing the space-creating properties of twentieth-century electroacoustic tools—what high-end audio calls "imaging"—they suggest that sound artists first mastered the equivalent of Renaissance linear perspective (Euclidian space) and then produced imaginative distortions that echoed the painterly discoveries of "Impressionism, Expressionism, Modernism, Cubism, Futurism, Dadaism, Surrealism."[32] What had taken painting centuries to accomplish was achieved in sound over the course of a few decades. Following the enabling example of Stockhausen, "modern audio engineers and electronic composers, without necessarily realizing their new role, became the aural architects of virtual, imaginary, and contradictory spaces."[33]

For their part, the Firesign Theatre had explicitly come to realize their new role by late 1968. Whereas, despite the accomplished absurdism of its title track, *Waiting for the Electrician* had still largely been a series of sketch performances ornamented with sound effects and music, a radical new set of diegetic space effects was apparent less than a minute into *How Can You Be in Two Places at Once*. An unnamed character (in the script he is called Babe) follows the siren call of a used car salesman across a busy street, onto the lot, and into a car. At the sound of a closing door, the ambient mayhem of the road—cars passing, bottles breaking, an unexplained marching band—disappears and a sudden wash of reverb on the two characters' voices suggests they have entered a space far larger than a car interior. Stereo panning moves the characters around this new contradictory space, as the car's many features are demonstrated. A faucet incongruously turns on and off. The characters then commence to try out the car's numerous electronic devices, which accrue one on top of another and rapidly transform what once seemed a quiet and roomy space into one that sounds (to put it mildly) very busy and claustrophobic:

- in the center of the stereo field, a television plays a used car advertisement and then breaks back to a classical Hollywood movie, complete with crowd noise and period soundtrack ("People of Alexandria! Who burned the library?"), which continues to play while

- on the left the manic pitch ("Super Box Number Time! In the fifteen hundred, on the 27–62!") of an AM DJ cuts in and out of a commercial, accompanied by whistles and kazoos, and hectically runs on while

- in the right of the sound field, a smugly narcoleptic FM announcer IDs the show's patrons and then cuts back to programming ("Thanks for the insurrection, and now back to our morning concert of afternoon showtime favorites, the Magic Bowl Movement from Symphony in C Minus by Johann Amadeus Matetsky"), which turns out to be an excruciating musique concrète cutup of Beethoven's 7th symphony, and is still in the air as

- a thin shortwave broadcast announces, in Spanish, an advertisement for the same car dealer that has also been sponsoring the other stations.

Each of these fictional transmissions had been accomplished by making a "reduction mix"—several recorded tracks bounced down to one, highly detailed, track. To the listener, each broadcast is distinguishable by the genre that is being parodied (i.e., by the group's writing and performance); by the placement of the submix in the stereo field; and by distinct EQ filters, each connoting a different "medium" (the filter for shortwave radio being the thinnest and grainiest).

Embedded somewhat obscurely is the joke that all four broadcasts have the same sponsor, Ralph Spoilsport of Ralph Spoilsport Motors, and that the sponsor is himself there in the car. But the more visceral experience of listening conveys the deeper point, which is about the plurality and the ubiquity of media transmissions. Each authentically grating in its own right, cumulatively they indicate a space that could hardly be credibly represented visually, and their simultaneity obliges listeners to choose which part they will try to decode. It is an allegory as well as a joke.

All of this takes about four minutes of a twenty-eight-minute album side, the remainder of which is hardly less involved. At the slam of a door, the acoustics immediately become warm and nonreverberant (the automotive space sound scholar Karin Bijsterveld calls the "acoustic cocoon"), and we remain in the car with the protagonist as he drives off.[34] He discovers that the car's climate control is able to conjure different total environments while paradoxically remaining, for now, a car: winter wonderland, spring fever, and the eventually chosen "tropical paradise," which is the album's first reference to Vietnam. After about nine minutes, the car will disappear entirely but not before it is the site of another astonishing contradictory soundscape. A series of road signs, dramatized verbally, approach and recede from the listener via volume swells and move across the sound field from center to right. It produces an uncanny effect of movement on the highway, even as one subset of these "signs" contradicts this sense of movement by staging a version of Zeno's paradox (Antelope Freeway one-half mile, one-quarter mile, one-eighth mile, one-sixteenth mile . . . one two-hundred and fifty-sixth mile . . .), a contradiction of space and time together.

ARCHAEOLOGY

On one hand, such depth of detail was the self-evident product of Eno's "in-studio composition." With a contract that traded mechanical royalties for unlimited time in the studio, the Firesign Theatre were able to write and record iteratively and deliberately. In a way that resembled Geoff Emerick's work with the Beatles, their new Columbia engineer, Bill Driml, was open to breaking with established protocols, collaborating with the artists, and experimenting with new sounds. And, as they well knew, they were also availed of the state-of-the-art affordances of the CBS Columbia Square studio, newly installed with eight-track recording equipment (fig. 7). Located at the corner of Sunset and Gower in Hollywood, Columbia Square was where the Byrds had recorded "Eight Miles High," "2-4-2 Fox Trot (The Lear Jet Song)," and "C.T.A.—102," and it was where Moby Grape had recorded their first album and where Brian Wilson had recorded vocals for the Beach Boys' "Good Vibrations" and *Pet Sounds*. These were all recordings that had experimented with quasi-diegetic "architectural" effects.[35] Namechecking "Yellow Submarine," David Crosby said, "If we can put anybody on a trip where they feel the same things that we felt going up in that airplane then we've succeeded."[36]

But in a way that was not true for the musicians, the new Firesign album also directly invoked the longer history of Columbia Square. Built in 1938, the International Style building had been overhauled by CBS in 1961, anticipating what an in-house journal called "the advent of stereophonic sound and its completely new process of recording."[37] It now complemented, and in some ways surpassed, the label's legendary 30th Street studio in New York. The opulent Columbia Square had not been originally designed for recording but was rather a state-of-the-art facility made for radio broadcasting. As Neil Verma has pointed out, radio studios were at that time typi-

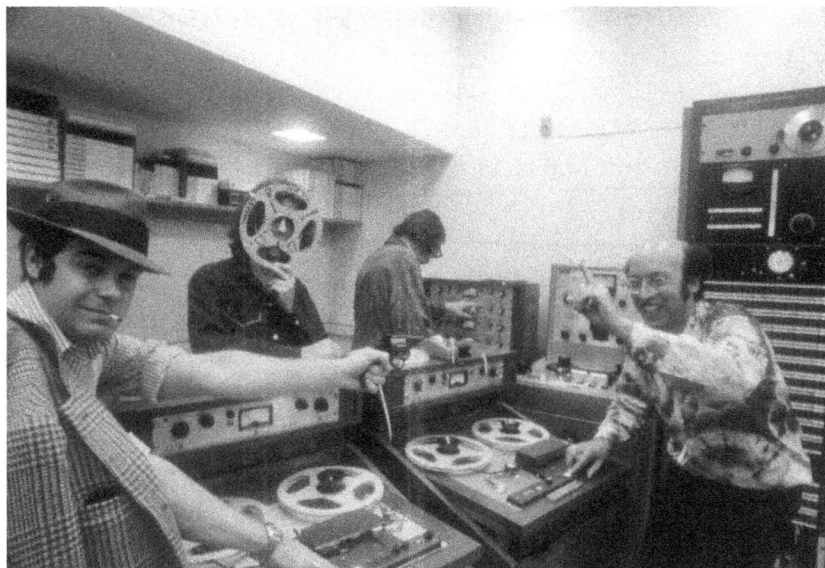

Figure 7. The Firesign Theater at CBS Columbia Square Studio, July 1969. Photo by Frank Lafitte. Courtesy of Sony Music Archives.

cally far more acoustically complex than those built for recording music.[38] For more than two decades, Columbia Square was home to fabled programs from Jack Benny, Burns and Allen, and *The Orson Welles Show* to *Suspense*, *Gunsmoke*, and *The Adventures of Philip Marlowe*.[39] The great playwright Norman Corwin regarded Columbia Square as radio's "Mecca. . . . There was not anything quite corresponding to its splendor in New York."[40]

With free rein to explore the premises, the Firesign Theatre discovered abandoned technologies and devices from the radio days throughout the building: there was the enormous echo chamber (which had since been used on musical recordings), a plethora of sound-effect devices for live Foley (doors, guns, wind machine, a board for footsteps), and the Hammond B3 organ made famous by *Suspense*. The most important discovery was a set of forsaken RCA ribbon microphones that were ideal for on- and off-mic voice-work. These gave *How Can You Be* a spatial depth that *Electrician* did not have and were used on all the group's subsequent Columbia albums.[41]

So while the mesmerizing Zeno's paradox/talking-road-sign sequence of *How Can You Be* required several up-to-the-minute technologies, not least the stereo movement abetted by the new eight-track tape machines, it also hearkened directly back to the eerie sound world of Lucille Fletcher's radio play "The Hitch-Hiker," which first broadcast from Columbia Square as an episode of *The Orson Welles Show* in November 1941. "The Hitch-Hiker" tells the story of Ronald Adams (Welles), a man driving across the country from Brooklyn to California. Setting out from his mother's house in Brooklyn, he passes a hitchhiker on the Brooklyn Bridge, sees him again on the Pulaski Skyway, and then continuously as he proceeds across the country—events Adams conveys, with increasing anxiety, entirely from the driver's seat of the car. The story becomes so deeply enmeshed with Adams's psychology that the effect of the play is, as Verma observes, to confuse the space of the outside world with a *state of mind*; the "acoustic cocoon" is transformed into a place of paranoid projection.[42] A final plot twist reveals that most of the story's events have occurred "outside of natural time."[43] In a way, these are all things that happen to *How Can You Be*'s protagonist, too.

Phil Proctor would later explain, "We wanted to produce the records as if radio had continued into the modern era with the full force of energy it had during its so-called golden age. What would it have sounded like?"[44] This is a succinct characterization of what would today be called media archaeology, a field of inquiry producing technical genealogies and countergenealogies, as well as what Thomas Elsaesser dubs a "poetics of obsolescence." Above all, its diverse strains are guided by "a strong sense/consensus that one should be 'doing media archaeology' rather than merely using it as a conceptual tool."[45]

Elsaesser also aligns media archaeology with the parallel development of "counterfactual history" (192), which could also be a dignified name for some of *How Can You Be*'s most notorious gags—spoiler alert: the president of the United States *is* named Schicklgruber—as well as a means of evoking another element of radio's "golden age" patrimony, namely, its historical entanglement with disinformation and propaganda. In the 1938–39 *Manual of German Radio*, Schicklgruber himself (I mean Hitler) wrote that "we should not have conquered Germany without . . . the loudspeaker."[46] And by that time, Dorothy Thompson, the first American journalist to be expelled from Nazi Germany, had already remarked that "radio is to propaganda what the airplane is to international warfare."[47] Nor was radio propaganda an exclusively fascist concern. While the United States remained officially neutral, British international broadcasting worked to cultivate feelings of solidarity and sympathy between the two countries, something that had been another object of study at the Princeton Listening Center.[48]

As the US entered the war, American commercial radio networks adapted their conventional genres and created specialized broadcasts to serve the war effort. The networks "willingly disseminated government propaganda and successfully united much of the American public behind the war effort."[49] And although many, including Welles, had contributed, the undisputed laureate of American antifascist radio had been Norman Corwin. "We Hold These Truths," his sesquicentennial celebration of the Bill of Rights, aired simultaneously on all four national networks in December 1941, eight days after Pearl Harbor. After many further programs, in May 1945, *On a Note of Triumph* celebrated the end of the war with another national broadcast; Columbia Masterworks released a six-disc recording (an album, in fact) before the end of the year. Both broadcasts had originated from Columbia Square in Los Angeles:

> So they've given up.
> They're finally done in, and the rat is dead in an alley back of the
>   Wilhelmstrasse.
> Take a bow, G.I.,
> Take a bow, little guy.
> The superman of tomorrow lies at the feet of you common men of this after-
>   noon.
> This is It, kid, this is The Day, all the way from Newburyport to Vladivostok.
> You had what it took and you gave it, and each of you has a hunk of rainbow
>   round your helmet.
> Seems like free men have done it again. . . .
> Far-flung ordinary men, unspectacular but free . . .[50]

The Firesign Theatre knew the pieces well, both from the disc recordings and from print collections of the scripts (which helpfully included Corwin's genial production notes), and they were drawn to his later experience as a writer blacklisted just three years after the war.[51] One vector for Phil Proctor's query—what would it have sounded like?—was therefore to wonder about the fate of Corwin's gregarious antifascism and its "mystical vision of citizenship" in the Vietnam era.[52]

Annoyed by his car's "tropical paradise" sound world, the protagonist of *How Can You Be* selects the climate control's "land of the pharaohs" as a means of escape. This choice elicits a pyramid, which may be in the Egyptian desert or on the back of a US dollar bill (or both). As the protagonist runs inside, the pyramid becomes "the only nice motel in town." Ostensibly, we may all still be in the car, but the car is never mentioned again. Instead, the motel's lobby becomes the site of an eight-minute Corwinian pageant in which the truisms of democratic citizenship are repurposed for the age of the imperialist war in Vietnam:

> BABE: [*singing*] This land is made of mountains
> This land is made of mud
> This land has lots of everything
> For me and Elmer Fudd
> This land has lots of trousers
> This land has lots of Mausers
> And pussy cats to eat them when the sun goes down . . .
>
> JOE: Stop!
>
> DESK CLERK: It wasn't always like that
>
> J: No, first they had to come from towns with strange names like . . .
>
> EDDIE: Smegma!
>
> DC: Spasmodic!
>
> E: Frog!
>
> J: And the far-flung Isles of Langerhans. . . .
>
> DC: And we took to them!
>
> E: And they took to us!
>
> DC: And what do you think they took?
>
> ALL: [*chanting*] Oil from Canada! Gold from Mexico!
> Geese from their neighbor's back yard! Boom, boom!
> Corn from the Indians! Tobacco from the Indians!
> Dakota from the Indians! New Jersey from the Indians!
> New Hampshire from the Indians! New England from the Indians!
> New Delhi from the Indians!
>
> B: Indonesia for the Indonesians!

J: [*cannon shot*] Yes, and Veterans' Day . . .

DC: But we couldn't do it alone [*Morse code sending under*]

J: No! We needed the Hope, the Faith, the Prayers, the Fears

DC: The Sweat, the Pain, the Boils, the Tears

J: The Broken Bones!

DC: The Broken Homes!

J: The total degradation of . . .

B: Who?

E: You! [*champagne cork pop and pour*]
   The Little Guy!

The pageant, as can easily be seen, mimics Corwin's rhetoric as well as his floridly extroverted style.[53] The "Little Guy" is a signature Corwinism, a stagey update of the Popular Front's "common man." Updating it for late 1968, the Firesign Theatre revealed the way Corwin's hortatory patriotism had been used to exploit the Little Guy while concealing the inegalitarian reality of the war (after first extending *Waiting for the Electrician*'s critique of Native American dispossession). When at length the sequence concludes, the protagonist finds himself cheerily coerced into enlisting in the army ("Get in step with the voices of the feet already dead in the service of their country!"), after which he appears briefly to become African American and then disappears from the story entirely—a set of sonic figures signifying the true demographics of the Vietnam War, where "in the early years of the fighting, blacks made up 23 percent of the fatalities."[54] David Ossman, who would later collaborate with Corwin on a fiftieth-anniversary broadcast of "We Hold These Truths," began his working notes for *How Can You Be* with the unfinished statement: "The problem with Norman Corwin . . ."[55] (The problem with the Firesign Theatre, meanwhile, was audible in their blackface answer to the problem of Corwin.)

With the protagonist now totally absent, the final seven minutes of the side involve a World War II vignette that concludes with a travesty of a USO-style singalong, which is titled, borrowing the SDS's slogan from Chicago 1968, "We're Bringing the War Back Home." This performance is then revealed to have been the conclusion of a film broadcast on the *Late Late Show for a Saturday Night*, which is followed by a channel-surfing

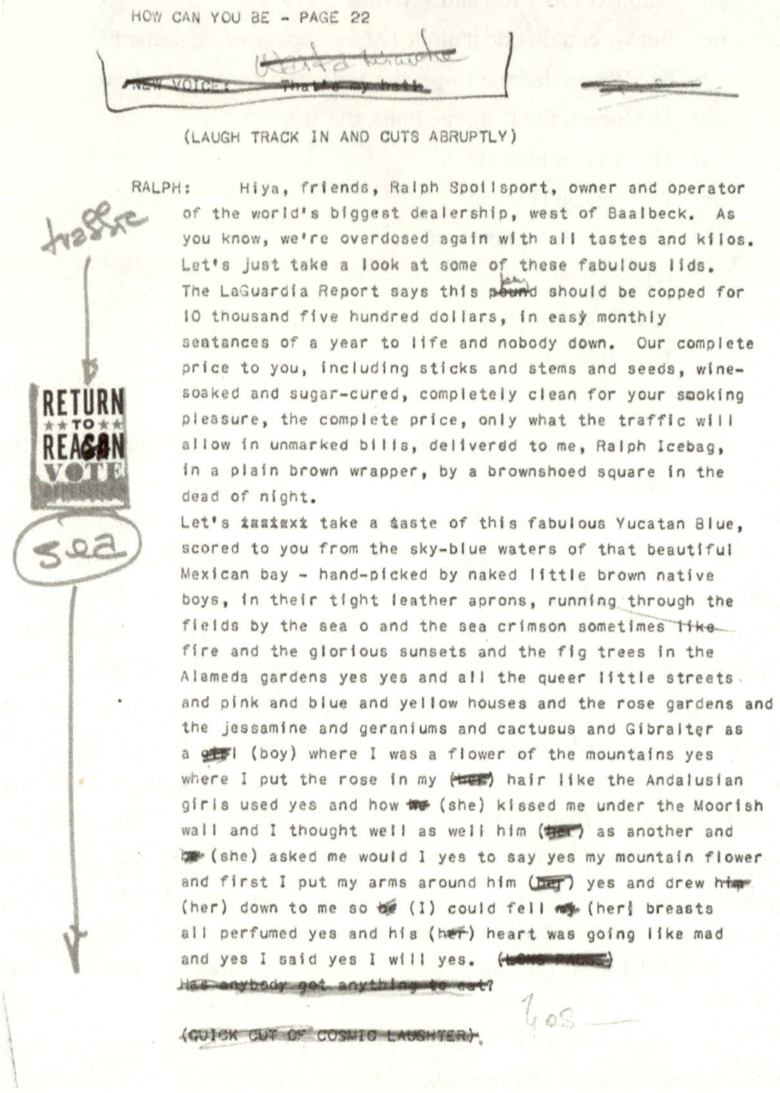

HOW CAN YOU BE - PAGE 22

(handwritten annotation)

~~NEW VOICE:~~ ~~That's my habit.~~

(LAUGH TRACK IN AND CUTS ABRUPTLY)

RALPH: Hiya, friends, Ralph Spoilsport, owner and operator
of the world's biggest dealership, west of Baalbeck. As
you know, we're overdosed again with all tastes and kilos.
Let's just take a look at some of these fabulous lids.
The LaGuardia Report says this ~~pound~~ should be copped for
10 thousand five hundred dollars, in easy monthly
seatances of a year to life and nobody down. Our complete
price to you, including sticks and stems and seeds, wine-
soaked and sugar-cured, completely clean for your smoking
pleasure, the complete price, only what the traffic will
allow in unmarked bills, deliverdd to me, Ralph Icebag,
in a plain brown wrapper, by a brownshoed square in the
dead of night.
Let's ~~instant~~ take a ~~t~~aste of this fabulous Yucatan Blue,
scored to you from the sky-blue waters of that beautiful
Mexican bay - hand-picked by naked little brown native
boys, in their tight leather aprons, running through the
fields by the sea o and the sea crimson sometimes ~~like~~
fire and the glorious sunsets and the fig trees in the
Alameda gardens yes yes and all the queer little streets
and pink and blue and yellow houses and the rose gardens and
the jessamine and geraniums and cactusus and Gibralter as
a ~~girl~~ (boy) where I was a flower of the mountains yes
where I put the rose in my ~~(her)~~ hair like the Andalusian
girls used yes and how ~~he~~ (she) kissed me under the Moorish
wall and I thought well as well him ~~(her)~~ as another and
~~her~~ (she) asked me would I yes to say yes my mountain flower
and first I put my arms around him ~~(her)~~ yes and drew ~~him~~
(her) down to me so ~~he~~ (I) could fell ~~my~~ (her) breasts
all perfumed yes and his (her) heart was going like mad
and yes I said yes I will yes. ~~(LONG PAUSE)~~
~~Has anybody got anything to eat?~~

~~(QUICK CUT OF COSMIC LAUGHTER)~~

(handwritten margin notes: "traffic" with arrow; "RETURN TO REAGAN VOTE" logo with "sea" circled and arrow; "408")

Figure 8. Working script from *How Can You Be in Two Places at Once When You're Not Anywhere at All*, showing production notes for ambient noise, morphing from "traffic" to "sea" as the speaker Ralph Spoilsport transforms into Molly Bloom from James Joyce's *Ulysses*. Firesign Theatre Collection, National Audio-Visual Conservation Center, Library of Congress.

sequence of TV broadcasts that return us at length to Ralph Spoilsport, now selling weed rather than cars, who seamlessly segues into the final 150 words of Molly Bloom's soliloquy from the end of James Joyce's *Ulysses* as the sound of cars on the highway morph into an oceanic chorus of "yeses," "Yes I will yes yes" (fig. 8). The car has long since disappeared, the protagonist has gone, and the listener is left with the euphoric sounds of a gender-ambiguous speaker's unconditional surrender. (At the end of an actual broadcast day in 1969 listeners would have heard the national anthem.)

DRILL AND DISTRACTION IN THE YELLOW
SUBMARINE (SLIGHT RETURN)

Of the many gnomic phrases coined by the German media theorist Friedrich Kittler, this may be the most famous: "The entertainment industry is, in any conceivable sense of the word, an abuse of army equipment."[56] Appearing in the 1986 book *Gramophone, Film, Typewriter*, the slogan appears as the moral of a (true) First World War story in which the new technology of wireless radio, theretofore reserved for military communications, had been used to relieve soldiers' sensory deprivation in the trenches of the Ardennes. When higher-ups learned that musical recordings and news items were being broadcast to troops, they banned the practice, calling it "an abuse of army equipment."

That experiment anticipated radio as we came to know it, and Kittler appropriates the CO's punitive phrase with a Firesignian irony. His argument, moreover, hews closely to the path disclosed by *How Can You Be in Two Places at Once*. Kittler goes on to observe that the technologies that would be rearticulated in the 1950s as *high fidelity*—frequency modulation and signal multiplexing (VHF and stereo)—were first designed for military communications and intelligence operations in the Second World War. A precursor to the hi-fi LP was used to train RAF Coastal Command in the detection of German submarines; full-frequency propaganda broadcasts were used as counterintelligence before the Battle of the Bulge; multiplex transmissions guided German air raids over Britain (and were the source of their eventual defeat) (99–103).

Taken together, Kittler suggests, these innovations enabled an unprecedented power for misdirecting or otherwise controlling human beings,
and the ominous implications did not disappear when the technology was
repurposed in consumer goods after the war. The World War II pilot was
"the first consumer of a headphone stereophony that today controls us all"
(99–100). "Ever since EMI introduced stereo records in 1957, people
caught between speakers or headphones have been as controllable as
bomber pilots" because "hi-fi stereophony can simulate any acoustic space,
from the real space inside a submarine to the psychedelic space inside the
brain itself. And [furthermore!] should locating that space either fail or
be a ruse designed to fool the consumer, it is only because the supervising
sound engineer has proceeded as shrewdly as the disinformation campaign prior to the Battle of the Bulge" (103).

No doubt Kittler would have extended his acid observations to include
Phil Austin (he of the 360th Psychological Warfare Unit) since he had
been the Firesign member most involved in the production of their
albums.[57] Nor was it any small irony that the Beatles' "Yellow Submarine,"
released on the EMI subsidiary Parlophone, should have sung "of the literal chain that linked Liverpool's submarine crews to postwar rock groups"
(103). That chain had already been forged by Decca Records as the
company pressed the training record for the RAF Coastal Command,
and it would be forged anew by the hi-fi equipment the enlightened
army command now happily made available to US soldiers deployed
in the fields of Vietnam.[58] The one possible redemption of such an
otherwise bleakly determined and violent gestalt, Kittler intimates, resides
precisely in the "psychedelic space inside the brain," and the Firesign
Theatre do not disagree. Having dispensed with its protagonist, as well
as (long since) the plot-motivating car, when *How Can You Be*'s troubling
journey finally dissolves into undulating binaural waves of Joycean *yeeeses*,
they are washed in the phasing effect made famous by the end of
Jimi Hendrix's "Axis: Bold as Love" in 1967 (he of the 101st Airborne).[59]
Corwin's "We Hold These Truths," by contrast, had ended by asking "Is
not our Bill of Rights more cherished now than ever? The blood more
zealous to preserve it whole?" and then cut to a White House message
from FDR.[60]

If "Yellow Submarine" hails the naval bases of Liverpool, Kittler implies, it is less for the lyrics Ringo sings than for the "impossible [hi-fi] space" that houses them. At once inside the submarine and the psychedelic brain, it is synonymous with Blesser and Salter's "contradictory space." Like the car on Ralph Spoilsport's lot, it is also a diegetic space peopled by ersatz sailors and their friends, who are all aboard: Brian Jones, Marianne Faithfull, Patti Boyd, Neil Aspinall. Making this impossible space still more vivid, they have liberated sound-effects devices from an Abbey Road supply room ("chains, ships' bells, handbells from wartime, tap dancing mats, whistles, hooters, wind machines, thunder-storm machines"), discoveries the Firesign Theatre would make two years later at Columbia Square.[61]

GOONS AND BEATLES

Conspicuously, though, Kittler has nothing to say about the most obvious fact about "Yellow Submarine," which is that it is funny (Christgau: "in addition to everything else, [the Beatles] were the funniest rock stars ever").[62] If he had, he might have been led to examine the connections between the put-on and propaganda, something that is relevant not only to the Firesign Theatre. "Yellow Submarine" is the first official release on which the Beatles openly paid homage to *The Goon Show*, the anarchic and wildly inventive radio program that aired throughout the 1950s on the BBC Home Service and starred Peter Sellers, Harry Secombe, and Spike Milligan, who also wrote the episodes. All of them had served in the Second World War; Milligan and Secombe met on active duty in the North African Campaign. Characterized by surreal humor, puns, sound effects, silly voices, and shaggy outré plot lines, *The Goon Show* was popular with, in John Lennon's words, "'eggheads' and 'the people.'"[63] It was received by London intellectuals in the same context as the first Beckett and Ionesco productions, with which it was contemporary; but it was also a mainstay of British adolescents right through Lennon's meeting Paul McCartney in 1957 (in the early years they performed together under a name borrowed from the Goons, the Nerk Twins).[64] When the Beatles learned that George Martin had produced records for Milligan and Sellers, they trusted him completely. A decade later, when Lennon reviewed *The Goon Show Scripts* for the *New York Times*, he suggested that Nixon should be sent a copy.[65]

The Goon Show had been a major influence on the Firesign Theatre as well, something they intuitively understood as connecting them further to the Beatles. Ossman and Austin had discovered the Goons in the mid-1950s, when selections from the BBC show were included as a part of NBC's eclectic weekend program *Monitor*. Austin was later tasked with programming old *Goon Shows* for broadcast on KPFK, at the same time that he was producing *Radio Free Oz* for Peter Bergman.[66] Bergman, for his part, had briefly worked with Milligan in London as he made his circuitous route from Berlin in 1964 to Los Angeles in the summer of 1966.

Somewhat like Austin, Spike Milligan had been trained in military communications (he was a signaler early in the war), and it is tempting to see the Goons' war experience as a touchstone for the Beatles and Firesign Theatre both.[67] In a good short discussion of Beatles and Goons, Jonathan Gould points to the origins of the Goon style in military service humor, which perhaps goes some way to explaining why the Beatles' style of psychedelia was at once military and comic (49).[68] Although unmentioned by Kittler, another notable feature of "Yellow Submarine" was what had been put in the traditional place of a solo: a tape salad of military brass band marches, cut and spliced by Emerick.[69] In retrospect, it is not hard to connect these sounds to the Day-Glo military band that would soon take the Beatles' place on the cover of *Sgt. Pepper*, time-traveling acid casualties of the Crimean War. Maybe this is what inspired Langdon Winner to write that "the closest Western Civilization has come to unity since the Congress of Vienna in 1815 was the week the *Sgt. Pepper* album was released."[70] (But that outrageous suggestion only obtains its full meaning when read together with Richard Goldstein's famous dissenting review: "By the third depressing hearing, it begins to sound like an immense put-on.")[71]

Gould goes on to observe that "Goon humor was cult humor, and the essential principle of cult humor can be stated as follows: The more obscure the joke, the greater the intimacy that comes from sharing in it." It was "a principle that Lennon grasped intuitively" (50). Gould is speaking of the teenage John, but when Lennon confided to Hunter Davies around the time of *Sgt. Pepper* that "we talk in code to each other as Beatles," he was no doubt describing that same intimacy, cultivated and protected by Goonian doublespeak.[72] One of the paradoxes of Beatlemania was that such intimacy should also have seemed so overtly evident. Yet their shared humor and affection was one of the things that seemed revolutionary; Allen Ginsberg remembered it as the first time anyone had seen men be affectionate with each other so publicly.[73]

The ur-location for these public/private performances were the Beatles' early to mid-period media interviews, the template for the dialogue of *A Hard Day's Night*. Kevin Eldon's description of these interviews, notably, sounds very like the Firesign Theatre's improvised radio programs—each Beatle behind a microphone "dodging questions, holding their own conversations, confusing their interrogators with non-sequiturs . . . their 'riffing,' picking up on each other's cues and commenting on and embellishing each other's contributions . . . a sort of spoken version of musical jamming."[74] Fans of the Firesign Theatre's radio programs heard a Beatlesque intimacy listening to the group's surreal, almost telepathic improvisations.[75] And because they were written to be listened to many times over, Firesign's albums fostered a parallel mode of sociability among listeners, who often memorized phrases or long sequences from the records involuntarily.[76] Like the Beatles' shared code, these passages could then be used in public to express secret affinity among initiates, provide ironic or nonlinear commentary, or indoctrinate new cognoscenti. The argot became a collective possession of the fans. And you can believe me because I never lie, and I'm always right. Take off your shoes for industry. I was right about the comet.

This transitive property of Firesign speech shows the way the late 1960s saw a mode of sociability, inaugurated through the Beatles, that extended the intimacy exemplified by cult humor to a broader cultural level. Greil Marcus realized this in a long early essay that described the 1960s' sense of "rock and roll as a secret and common experience." "To initiate an outsider into our scheme of things," he wrote, "was to initiate him into the Beatles; not the Beatles as legitimate music, but the Beatles as the Beatles—as noise, humor, fun, sound, bite, tradition, hair, and cynicism." Anticipating his regnant phase, he affirms that "we began to understand it as a kind of culturally secret parallel history for a community that recognized itself as such only through the rock."[77]

Writing for *Creem* in 1971, Marcus was seeing signs of that community's fragmentation, and that fragmentation was the occasion of his essay. But Marcus had two years earlier initiated a somewhat bad-faith test of the community's limits as the author of one of the era's signature put-ons. Together with Bruce Miroff, Marcus had placed an ersatz review in the October 18, 1969, *Rolling Stone*. Appearing under the byline of "T. M. Christian"—a pseudonym taken from a new movie, *The Magic Christian*, starring a Beatle and a Goon (Ringo and Peter Sellers)—it announced the existence of the Masked Marauders, a purported super-group comprising three Beatles, Bob Dylan, Mick Jagger, and an unknown drummer (secretly recorded in Canada by Al Kooper!). Introduced by "rumors of an event that at first seemed hardly believable," T. M. Christian opined that "this album is more than a way of life; it *is* life."[78] *San Francisco Chronicle* columnist and *Rolling Stone* mentor Ralph J. Gleason quickly identified it as a put-on. But it was already too late; the T. M. Christian review had caused such excitement that an actual album was recorded to correspond with the review's invented details, commissioned and released on the fabricated Reprise subsidiary Deity. (It's now available on all the streaming services.)

The range of media involved—magazine, newspaper, FM radio, LP album, more distantly cinema—clearly places the Masked Marauders experiment in the pantheon of late 1960s put-ons. But a closer examination reveals something more about its relation both to the "secret and common experience" (a descendant of the Beatles' "code") that Marcus would theorize in *Creem* and to propaganda. For whether Marcus realized it or not, the prank expressed his desire as well as his disdain; it mocked the economy of rock celebrity while also expressing some of his deepest wishes for music. Marcus would later describe Beatlemania unironically in terms nearly identical to those with which he had promoted the fictional Masked Marauders record: "Excitement wasn't in the air; it was the air."[79] The Masked Marauders put-on hewed the road between a playful and speculative secret language and actual critical knowledge (or *gnosis*), a technique he would perfect in his book dedicated to the Firesign Theatre, *Lipstick Traces*.

A cognate scenario could be found in the way the intensively realized worlds of Beatles albums post–*Rubber Soul* were, at the very moment of the *Masked Marauders* put-on, being newly construed as fields of clues for decoding in accordance with the hermeneutic travesty of the Paul-is-dead hoax (the walrus was Paul; *walrus* is the Greek word for corpse; Paul is wearing a black carnation in *Magical Mystery Tour*, etc.). It can be briefly summarized as follows: Paul McCartney was supposed to have died in a car accident in 1966 and replaced by a look-alike according to a secret arrangement masterminded by John Lennon but agreed on by all the Beatles in order to (a) start a new religion or (b) profit through sustained and increased record sales. The Beatles then peppered their subsequent work (lyrics, album art, record production tricks like backmasking) with clues that could reveal the entire story to fans learned and practiced enough to uncover them. Between September 17 and November 30, 1969, Paul-is-dead grew from a modest piece in the student newspaper of Iowa's Drake University (2024 undergraduate population: 4,884) to an international cause célèbre culminating in a *Life* magazine cover story, a special investigative television broadcast hosted by celebrity lawyer F. Lee Bailey, and countless other artifacts.[80]

Especially noteworthy were both the highly participatory quality of the hoax and the unstable quality of belief that it both generated and exploited. While some were undoubtedly true believers, others took for granted that it was a put-on but were nevertheless sure that the Beatles were leaving their audience clues to decode (in one theory, the Beatles had bought an island and were disclosing its location throughout their songs).[81] In November, a column in the *Los Angeles Free Press* debunked Paul-is-dead while nevertheless affirming "all the clues that the Beatles purposely have strewn around over the years."[82] As the resemblance here to QAnon suggests, the participatory aspect of Paul-is-dead also resembled a key dynamic of propaganda, which was to inspire a falsehood's growth and proliferation with the cooperation of its targeted audience.[83] The Detroit DJ who had been instrumental in making Paul-is-dead go viral, Russ Gibb, urgently tried to contact T. M. Christian, believing the Masked Marauders were further confirmation of McCartney's death.[84]

The horrific counterpart of Paul-is-dead—its double and, as Devin McKinney has recognized, its other exact contemporary—was the spate of murders committed by the followers of Charles Manson in Los Angeles, August 1969. Legendarily, these, too, were inspired by the Beatles, though Manson's tendentious misreadings from the White Album and "Yellow Submarine" would not be known until early 1970.[85] In its official reception, 1968's White Album (*The Beatles*) had itself been understood as a put-on, albeit in a different way from *Sgt. Pepper*; instead of a fictive world, the White Album's put-ons consisted of its genre pastiches and open self-referentiality (here's another clue for you all).[86] Manson heard it as propaganda: "He juggled their words to suit him so that the Beatles' 'Revolution' turned into a strange kind of Mein Kampf."[87] In a more muted register, the Masked Marauders put-on had functioned like propaganda, too: the review had elicited phone calls to FM stations across the country demanding the album, which was then dutifully recorded and released.[88]

Why this particular rabbit hole? Because, as any casual Firesign Theatre fan knows, *How Can You Be in Two Places at Once When You're Not Anywhere at All*—which peaked on the *Billboard* charts the week ending October 25, 1969—is famous for containing as many as two dozen Beatles references. These ranged from the self-advertising "All Hail Marx and Lennon" cover image and numerous quotations (everyone knew her as Nancy, my story had more holes in it than the Albert Hall, goo goo g'joob, sir!) to blackmail artist "Rocky Rococo" and the psychedelic return of Ralph Spoilsport at the end of side 1 (which nodded to the reprise of *Sgt. Pepper*'s title track). *Creem*'s grimly admiring review of *How Can You Be* concluded by suggesting that if you held the album under an infrared light you could see "Russ Gibb molesting Paul McCartney." ("But you know where rumors are at.")[89]

FIVE YEARS ON THE FM BAND

The Buffalo radio station WKBW broadcast "Paul McCartney Is Alive and Well (Maybe)" at the height of the Paul-is-dead craze on Halloween night, 1969. The seventy-minute program heatedly collated all the "clues" that told of how Paul had blown his mind out in a car on a stupid bloody Tuesday in 1966 and explained John's ensuing conspiracy to cover up and cash in (also buy an island, start a new religion, etc.). Exactly a year earlier, the same station had staged an anniversary adaptation of the Mercury Theatre's *War of the Worlds*.

The Buffalo 1968 *War of the Worlds* was a surprisingly lavish production that could boast one dimension of realism the original had lacked. Using the familiar voices of KBW's regular hosts and reporters, the fiction fit seamlessly (which is to say ambiguously) in and out of the usual programming stream. KBW's Total News Department led its 11:00 p.m. summary with actual news: LBJ's announcement of a bombing halt in North Vietnam and promise to extend peace negotiations to the Vietcong (an October surprise already covertly sabotaged by candidate Nixon).[90] It ended three minutes later with reports of "huge explosions that have been taking place on the planet Mars." Returning to its usual programming—Cream's "White Room," the Beatles' "Hey Jude," Buffy Sainte-Marie's "I'm Gonna Be a Country Girl Again"—subsequent bulletins interrupted to report the Martians' landing on Grand Island and progress into Canada and the Finger Lakes region.[91] Taken together, the 1968 (Martian) and 1969 (McCartney) broadcasts demonstrate the abiding hospitability of radio for the dissemination of put-ons and disinformation. Though each at some point announced its fictionality—its official denial—neither failed to cultivate the audience's credulity.[92] The Buffalo *War of the Worlds* seems even to have resurrected some of the hysteria that followed the 1938 broadcast.[93]

Even viewed without this Kittlerian eye, the adjacency of the two KBW broadcasts reveals something surprising—namely, the affinity, or even continuity, of golden age radio practices with the age of psychedelia and countercultural radio. This is a leitmotif, too, repeatedly found in the memoirs of DJs from the period, which discover the authors reviving the rapt engagement associated with the "theater of the mind" of the 1930s and 1940s.[94] A May 1968 advertisement for San Francisco's KSAN went out of its way to play with the idea (fig. 9).[95]

But nothing characterized this association more than *How Can You Be in Two Places at Once When You're Not Anywhere at All*, which, in addition to citing Norman Corwin and Lucille Fletcher, was structured as a mirror image of Welles's *War of the Worlds*. Welles's play begins as a simulated evening of radio that becomes interrupted by emergency bulletins and eventually cedes control of its airwaves to the government; after forty minutes, the audience is finally given the consolation of a protagonist. "How Can You Be in Two Places at Once" begins with a protagonist whom it progressively diminishes and eventually abandons; the piece ends with

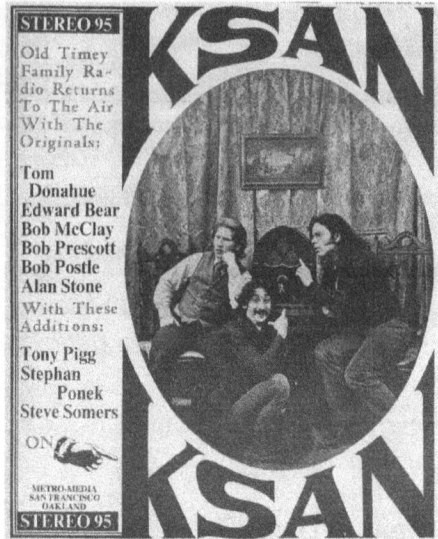

Figure 9. Firesign Theatre knockoffs Congress of Wonders posed around a vintage 1930s radio in an advertisement for San Francisco's KSAN. *Berkeley Barb*, May 31, 1968. *Berkeley Barb* courtesy of Raquel Scherr. All rights reserved.

a series of abruptly interrupted media transmissions—television channels surfed by an anonymous, disembodied viewer.

The structure of *How Can You Be*'s title cut was equally informed by "freeform" broadcasting, a relatively new technique originating on the FM dial and associated with stations like KSAN. Freeform emerged in the early 1960s on low-wattage noncommercial FM stations like New Jersey's WFMU and New York's WBAI, which came to define themselves in opposition to the strictly formatted playlists, frequent breaks for advertisements, and aggressively genial announcers of Top 40 AM (cf. "Big Box Number Time!"). The new FM style was characterized instead by long uninterrupted sets of a very wide range of sounds and by a much more laid-back or idiosyncratic presentation. These elements were exemplified by Bob Fass's *Radio Unnameable*, often cited as the first freeform program, which began its long run on WBAI, the Pacifica station where David Ossman had worked, in 1963.[96] Fass's overnight show featured live music, news reports, taped sound collages, poetry, free-associative rambling, audience call-ins, and numerous forms of music, all of which he orchestrated thematically. At once a seamless twenty-eight-minute piece and a series of freely associated episodes, *How Can You Be*'s title piece would undoubtedly have been heard in the context of freeform FM and played on its stations.[97]

There were good reasons Firesign felt at home with this technique. Peter Bergman's *Radio Free Oz* had clearly been inspired by *Radio Unnameable*, and when it began broadcasting in the summer of 1966, it was given the same overnight slot on Pacifica's KPFK that Fass occupied on its New York sister station. By the spring of 1967, the show had become such a sensation that Bergman had taken *Radio Free Oz* to the top-rated station in Los Angeles, KRLA-AM, beginning the night of the Easter Sunday Love-In. At KRLA the Wizard of Oz—with regular guests Ossman, Proctor, and Austin—enjoyed a once-a-week prime-time slot, more money, and a broadcast range that covered the entire LA basin, far larger than the feeble reach of KPFK.

Evidence of this moment's liminality is preserved in a KRLA aircheck that survives from April 1967. There, the paradigmatically chipper AM voice of Johnny Hayes can be heard announcing "two minutes past six, KRLA Big Time!" just before a promotional cart for *Radio Free Oz* collages together an air-raid siren, Indian raga, Sieg Heils from a Nazi rally, the Marine Corps Hymn, and Bergman intoning, "Radio Free Oz ... easy radio for troubled times, holy one."[98] That same week, the house publication *KRLA Beat* gamely affirmed that, with *Radio Free Oz*, "taking on Bergman's 'living trip' is truly a broadcasting innovation of the first water."[99]

By the summer, *Radio Free Oz*, preceded by a psychedelic program hosted by the Beatles' US publicist Derek Taylor, began broadcasting live before an audience at the Magic Mushroom club on South Ventura Boulevard. Here they interviewed luminaries like Cass Elliot and the Buffalo Springfield and reserved time each week for the performance of a new Firesign Theatre play, the writing and technique inspired by what they knew from the Goons' work on the BBC. With *Waiting for the Electrician* in production at Columbia Square, it was a crash course in writing in long form, parts of several scripts later ending up on the group's albums.[100]

But just as Bergman was leaving the low-wattage FM station for the big market of KRLA, FM had begun to undergo an institutional change that would expand the freeform technique more radically and less ambivalently than could have been the case at KRLA. For well over a decade, the majority of FM frequencies had been subject to the proprietary control of an AM parent station, which merely broadcast a duplicate signal on the corresponding FM channel (independent FM stations not on Pacifica were typically formatted for classical or "beautiful" music). In 1965, however, the FCC announced a new "nonduplication" policy mandating original programming for FM stations in the one hundred largest markets in the US. Dozens of frequencies would soon be made available for low-cost investment and experimentation on the FM dial, which, though it had a limited range, allowed for high-fidelity stereo transmissions, making it more amenable to the listening culture associated with the album. The new FCC policy became effective January 1, 1967, just in time for the so-called Summer of Love.

Between 1967 and 1969, as many as five dozen "underground" FM stations began broadcasting nationwide, many using the freeform style.[101] While their general eclecticism and left orientation resembled the Pacifica model, the new FM stations differed in key respects. Most obviously, they were commercial (though many at first made their own ads and operated on tiny budgets); they were also more specifically oriented toward music as opposed to the more theatrical eclecticism of *Radio Unnameable* or *Radio Free Oz*. Though this phase of radio broadcasting has been the subject of ingenuous hagiography, the most justly celebrated of these stations— KMPX (SF), KPPC (Pasadena), WABX (Detroit), WHFS (DC), WBCN (Boston)—militantly defended playlists that included jazz, blues, classical, and non-Western music alongside rock cuts and were totally amenable to playing entire album sides or very long tracks, as well as poetry and field recordings of birds. Many, moreover, also served as sources of political information of a kind that is almost totally absent from the airwaves today. As early as 1968, however, conglomerates such as Metromedia became major players in the FM market, heralding a compromise between commercial viability and artistic freedom.[102] Thus began a process that, imposing discipline and restrictions on programming and presentation, would culminate in the programmatically revanchist, white and male, mid-1970s format of album-oriented rock (AOR), and endured for decades.[103]

The Firesign Theatre's radio career rode these waves pretty precisely. They were fired from KRLA in the summer of 1968; early in 1969 they were fired again, this time from Metromedia's KMET-FM.[104] They came back a year later on the independent freeform station KPPC, and in the fall of 1970, they returned home to KPFK with a program syndicated to the Pacifica affiliates and sold via transcription discs to college stations, a formula they would repeat the following year (see table 1).

More revealing than Firesign's radio career, however, is the history of their albums' airplay. At the beginning of 1968, Phoenix's KNIX-FM was a "beautiful music" station; in May, the guitarist Buck Owens bought the station and converted it to country and western; by August KNIX had changed again, this time to a commercial underground format. That November, a *Billboard* magazine survey reported that KNIX's "Biggest Happenings" were a track from Hendrix's *Electric Ladyland* and Biff Rose's countercultural police satire "Buzz the Fuzz." Its "Biggest Leftfield Happening" was the Firesign Theatre's *Waiting for the Electrician or Someone Like Him*.[105] Four years later, in the summer of 1972, two DJs from Detroit's once freewheeling and politically radical WABX quit on the air (fig. 10). They cited the bad politics of the station's new sponsors ("we're selling too much death"), as well as the steady creep of the format that would become regularized as AOR: "It's just play more popular music and don't get quite so far out. And don't play jazz and don't play blues and don't play the Firesign Theatre."[106]

Together, these two stories are another allegory of the FM period, one that emphasizes the way the corporatization and regularization of the format was about something more than music. And it is not a selective example. Until 1969, WHFS in Washington, DC, had been a beautiful music station before a young DJ collective calling itself Spiritus Cheese (Sara Voss, Mark Gorbulew, and Joshua Brooks) brought their freeform show to the station, which led to the station's wholesale remaking. In April 1970, the DJs were fired for playing an aircheck of a recent Firesign Theatre concert.[107] (In the spirit of full disclosure, Spiritus Cheese had also hosted a Paul-is-dead news conference in late 1969.)[108] In 1972, John Neschi, a DJ at WOWI in Norfolk Virginia was first fired and then indicted for playing the Firesign Theatre and the Country Joe "fish cheer"; the *Chicago Seed* reported, "Oh yes. The DJ in question also had the habit of broadcasting the license numbers of unmarked police cars."[109] With its strong ties to the underground press and antiwar movement, Spiritus Cheese

Table 1 Firesign Theatre on Radio

Period	Program	Time slot	Station
July 1966–Oct. 1966	*Radio Free Oz* (Peter Bergman & Paul Jay Robbins)	Sunday–Thursday midnight–4:00 a.m., then midnight–3:00 a.m.	KPFK—90.7 FM Pacifica
Oct. 1966– March 1967	*Radio Free Oz* (Bergman solo)	Sunday–Thursday midnight–3:00 a.m.	KPFK
March 1967– Oct. 1967	*Radio Free Oz*	Sunday 8:00 p.m.–midnight	KRLA—1110 AM
Oct. 1967– Jan. 1968	*Radio Free Oz* *Live from the Magic Mushroom*	Sunday 9:00 p.m.–midnight	KRLA
Summer 1968	*Radio Free Oz*	Sunday 9:00 p.m.–midnight	KRLA
Nov. 1968– Feb. 1969	*Radio Free Oz* (Bergman and Ossman)	Sunday 9:00 a.m.–noon	KMET—94.7 FM Metromedia commercial "progressive"
Jan. 1970– July 1970	*Radio Hour Hour* (Firesign Theatre)	Sunday 8:00 p.m.–10:00 p.m. 6:00 p.m.–8:00 p.m.	KPPC—106.7 FM Independent commercial freeform
Sept. 1970– Feb. 1971	*Dear Friends*	Sunday 8:00 p.m.–9:00 p.m.	KPFK Pacifica + syndication
Nov. 1971– March 1972	*Let's Eat*	Thursday 8:00 p.m.–9:00 p.m.	KPFK Pacifica + syndication
Oct. 2001– Aug. 2002	*Fools in Space*	Saturday 6:00 p.m.–8:00 p.m.	XM Satellite Radio

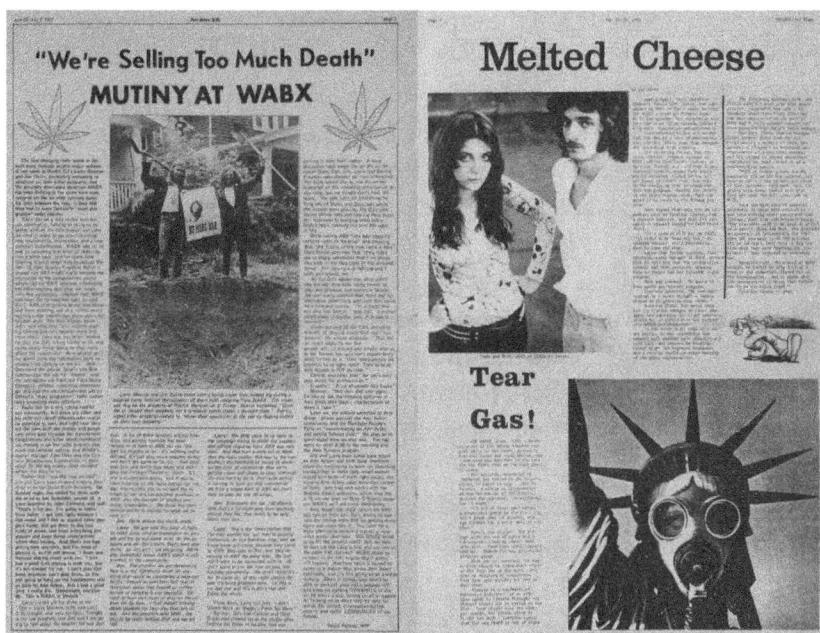

Figure 10. DJs disciplined for their radio broadcasts of the Firesign Theatre: Larry Monroe and Jim Dulzo, who resigned in protest from Detroit's WABX in 1972 (reported in the *Ann Arbor Sun*), and Sara Voss and Mark Gorbulew of the Spiritus Cheese collective, fired by the Washington, DC, station WHFS in April 1970 (reported in *Quicksilver Times*). *Ann Arbor Sun* courtesy of John Sinclair; *Quicksilver Times* placed in the public domain by the Sixth Column Collective.

(Voss in particular) were also known for "meticulously researched" thematic broadcasts that were often "risky and political."[110]

The pressure on the commercial FM stations was not only from the corporate bookkeepers; it was also coming from the FCC, which Nixon came to use as an instrument of political power through its selective implementation of the "Fairness Doctrine." At San Francisco's KSAN, the Black DJ Roland Young was compelled to resign for broadcasting a message from the Black Panther David Hilliard; this event was followed soon after by the forced departure of satirical tape collagist newsman Scoop Nisker.[111] As the music writers Steve Chapple and Reebee Garofalo would later remark, "It's ironic, but not atypical that the final solutions for progressive FM radio should be found with liberal corporate sponsors who eliminate all the politics of the experimental FMs, but keep the music and record ads." They point out that the October 1971 "mass political firings at KPPC," which followed the installation of a new transmitter greatly extending the station's range, were "directed particularly at the Credibility Gap."[112] Specializing in satirical newscasts, the Credibility Gap was a comedy group that had come to KRLA a year after *Radio Free Oz* and later went to KPPC.[113]

Justly weary of rockist hagiography, critics and scholars of popular music have often written against nostalgic accounts of FM's heyday. Viewed from the perspective of the total field of a station's programming—in which music is only one element—it is easier to see that the changes that came to FM were not the result only of commercial (or aesthetic) decisions. Seen institutionally, it makes less difference whether the Firesign Theatre was itself viewed as political or was used as a wedge to depoliticize FM radio. By 1977, Chapple and Garofalo could state plainly that "the political promise of the underground radio is largely a joke," while making one significant exception: Pacifica stations like KPFK.

OUR RENDEZVOUS WITH DESTINY IS TO UNCONDITIONALLY SURRENDER

The first side of *How Can You Be* had excavated the legacy of wartime radio inside the Columbia Square studio and on the FM dial. The album's second side pursued a much more frankly literal mode of media archaeology. "The Further Adventures of Nick Danger" had been pitched to Metromedia's KMET as part of a new FM show Firesign would broadcast live from the majestic Elks Hall abutting MacArthur Park.[114] When KMET turned them down, the Firesign Theatre recorded it on the soundstage of Columbia Square's Studio B, learning-by-doing the techniques of period broadcasts at the site of former CBS detective shows like *Yours Truly, Johnny Dollar*, and *The Adventures of Philip Marlowe*. Whereas *How Can You Be*'s title piece featured a dizzying array of edits and overdubs, "Nick Danger" was tracked entirely live (fig. 11). The group performed it together, producing their own sound effects and ambient sound with their friend David Grimm accompanying on the house Hammond B3 organ,

Figure 11. The Firesign Theatre recording "The Further Adventures of Nick Danger" live in Columbia Square Studio B, February 1969. Phil Proctor (*third from left*) is using a vintage Foley technique to create the sound of footsteps. Firesign Theatre Collection, National Audio-Visual Conservation Center, Library of Congress.

the same instrument that had accompanied the dramas of *Suspense* decades earlier.

For almost a half hour, "Nick Danger" plays as a relatively straight parody, faithfully employing golden-age studio equipment and technique. The plot is a storehouse of hoary narrative conventions including an extortionist, woman-with-a-past, a decadent mansion, and broadcast/production conventions like interior monologues and flashbacks aurally differentiated by high-pass filters. Phil Austin's Nick Danger weaves back and forth between narrator and character, sometimes within a single sentence, just as Philip Marlowe had been written for the actor Gerald Mohr.[115] Ossman is a totally convincing emcee and a geriatric butler; Bergman is a corrupt cop; Proctor's Rocky Rococo riffs on Peter Lorre's Joel Cairo from *The Maltese Falcon*. Though spiked with anachronistic references to drugs, the Beatles, and the occult, the pastiche remains formally faithful until four minutes from the end. At this point the group showily introduces a contemporary studio technique—the overdub—to blow apart the piece's narrative agreement and impose an audacious nonconclusion: a put-on that will pull together the first side's citations of the Vietnam and Second World Wars, and their interest in radio propaganda.

An accident involving a time machine, narrated from within a flashback, results in the doubling of all the story's principal characters, as the flashback attempts to return to the "present." The protagonist/narrator Nick Danger seeks to return the scene to diegetic normativity:

NICK DANGER: (VOICE DOUBLED) I did a quick twenty-twenty on the whole scene. I had thought that I was the only person going insane, but now we were all in this together. I knew what I had to do. I didn't like it, but that had never stopped me before. Alright, everybody! Take off your . . .

(VOICE BREAKS OFF ABRUPTLY)

ANNOUNCER: Ladies and gentlemen—we interrupt this scheduled transmission to bring you a message of national importance from the White House in Washington, D.C. . . . Ladies and Gentlemen: the President of the United States . . .

PRESIDENT: My fellow Americans: This morning, at 6:25 A.M., Pacific Standard Time, combined elements of the Imperial Japanese navy and air force ruthlessly attacked our naval base at Pearl Harbor in the Hawaiian Islands. I have conferred this morning with the Congress and the Chiefs of Staff in emergency session. We have reached our rendezvous with destiny! It is our unanimous and irrevocable decision that the United States of America *unconditionally surrender*! And now, my wife and I would like to return with you for the thrilling conclusion of "Private Nick Danger, Third Eye."

NICK: (CUTTING BACK IN) . . . guess I've solved another one for you.

SGT. BRADSHAW: Danger, I'll never know how you do it. I was sure I had the goods on ya this time!

NICK: Well Bradshaw—it's like in the Army, you know—"The Great Prince issues commands, founds states, vests families with fiefs. Inferior people should not be employed."

BRADSHAW: Nick, I can't knock success, but you still put me through too many changes.

As throughout "Nick Danger," the most obvious referent here is to radio's golden age, the overt pastiche of the radio detective genre and FDR's counterfactual Pearl Harbor surrender announcement each evoking radio's preeminence during the Second World War.

The explicit staging of an emergency announcement from the White House, moreover, paid homage to the interruptive style of the first half hour of Orson Welles's *War of the Worlds*, which was itself bound to the events of World War II. Kittler has called Welles's broadcast "sheer prophecy, feedforward of World War II," thinking of the way the United States would mobilize the media and its citizens in the service of the war effort, and he is also thinking of the simulated attack on a radio station—the "Gleiwitz Incident"—that the Nazis would use as their pretext for invading Poland, and starting the War, in September 1939.[116] In fact, *The War of the Worlds* was not as autonomously proleptic as Kittler suggests. CBS broadcast Welles's drama just a month after the Munich Accord, and it was in the tense weeks that led up to that infamous act of appeasement that the "breaking news" technique had been invented.[117] Dorothy Thompson, who in 1934 had seen the way the failed Austrian "July Putsch" had hinged on the takeover of the national radio service, viewed *The War of the Worlds* as a canary in the coal mine for Americans' susceptibility to propaganda. The play threw "more light on recent events in Europe leading to the Munich pact than everything that has been said on the subject by all the journalists and commentators. . . . If people can be frightened out of their wits by mythical men from Mars, they can be frightened into fanaticism by fear of Reds, or convinced that America is in the hands of sixty families, or aroused to revenge against any minority, or terrorized into subservience because of any imaginable menace."[118]

Meanwhile, far more available to the Firesign Theatre's contemporary audience was the way Nick Danger's final sequence incongruously employed a hallmark technique of late-1960s freeform radio, which, especially in California, indulged in manifold trendy spiritualist practices.[119] These ranged from reading horoscopes and tarot cards to, in this case, consulting the *I Ching*, the source of Nick Danger's hilariously arbitrary response to Sgt. Bradshaw: "The Great Prince issues commands, founds states, vests families with fiefs. Inferior people should not be employed."[120] Bergman regularly used new age practices on *Radio Free Oz*, and the group has always acknowledged that their use of the *I Ching* was not only satirical. Though they had tracked "Nick Danger" live, they had done so without having written an ending. In this context, the *I Ching* had comic value as a non sequitur but was also being employed as an aleatory compositional device, something Firesign knew the composer John Cage had done a decade earlier.[121] When they threw the hexagram for "The Army" together with a changing line leading to the hexagram "Youthful Folly," they felt especially vindicated, since the first side of the album had concluded its coercive Corwinesque pageant by dispatching their naive protagonist into the army.[122]

Though historically distinct, it is the unstable multivalence of propaganda that conjoins the FDR and freeform references. The counterfactual put-on of Roosevelt's "unconditional surrender" broadcast, though quite evidently a joke, is funny because its content is disinformation. To raise the question of US surrender in early 1969, however, was unmistakably also to express the desire for an immediate end to the war in Vietnam and in this way could be heard in the context of the kind of openly dissenting broadcasts on Pacifica, KSAN, or WABX. Together they expressed both the Firesign Theatre's wishful "feedforward" for the end of the war and an anxiety about its relationship to the last world war.

Four years later, John Lennon and Yoko Ono would stage a put-on that reprised, unconsciously or not, the Firesign Theatre's feedforward of unconditional surrender. In response to Nixon's attempts to have the peace activists deported, Lennon and Ono held a press conference in which they declared themselves ambassadors of the newly founded nation of Nutopia. They requested recognition from the United Nations and diplomatic immunity; then, "with a characteristic instinct for showmanship, they each whipped out a white tissue and said, 'This is the flag of Nutopia—we surrender to peace and to love.'"[123] In footage of the press conference, Lennon can be seen wearing a Firesign Theatre button on his lapel (fig. 12).[124]

To invoke FDR as the subject of disinformation in the same breath as an imagined post–Pearl Harbor surrender, however, was also to gesture toward a history of isolationist propaganda. For this reading, Exhibit A would be *The Truth about Pearl Harbor*, a 1945 booklet from the Nazi-adjacent isolationist group "America First," which alleged that FDR had goaded the Japanese attack as pretext for the US to enter the war.[125] To cite the FDR "advance knowledge" conspiracy theory in the present context of Vietnam was to evoke its structural similarity to the dubiously

Figure 12. John Lennon wearing a Firesign Theatre pin ("Not Insane") with Yoko Ono at the Nutopia press conference, April 1, 1973. Bettman via Getty Images.

alleged (now discredited) attack on the destroyer USS *Maddox* that was used to justify the Gulf of Tonkin Resolution of August 1964, authorizing the first American troop deployments in South Vietnam.[126] Though it was less likely the deliberate act of PSYOPs operations, the Gulf of Tonkin incident served as the Gleiwitz of Vietnam.

If the *I Ching* invoked spiritual dualities—referenced punningly in *How Can You Be*'s doubling of characters, as well as in the two distinct sides of the album—propaganda showed duplicity to be the double of duality. In liner notes written for the 1988 Mobile Fidelity Sound Lab reissue of the album, Austin discusses the album's formal and spiritual dualism at length.[127] Ossman's companion essay darkly emphasizes the Vietnam context: "Winning hearts and minds. Strategic hamlets. Body bags."[128]

Firesign's counterfactual FDR broadcast proved generative in a broader sense, providing a context that conjoined both sides of the album (as well as the next two), all of which could be seen as taking place in a United States that had secretly succumbed to fascism on or around World War II:

MORRIE: Sit down, Lurlene, this may be rough. The president of the United States *is* named Schicklgruber.

LURLENE: I'm going out there, Morrie. Help me into this parachute! . . . I'm going out there because I'm Bringing the War Back Home!"

This vision resonated with a growing consensus both about the war abroad and the police and National Guard at home, as when from the convention stage Senator Abraham Ribicoff decried the "Gestapo tactics in the streets of Chicago" in August 1968.[129] And it also channeled the history of *psychological warfare*, a term that entered the American lexicon in 1941 as the newly founded Office of Strategic Services (forerunner to the CIA) "viewed an understanding of Nazi psychological tactics as a vital source of ideas for 'Americanized' versions of many of the same stratagems."[130]

This point would be of more than anecdotal interest to someone trained in PSYOPs during the Vietnam years, as Phil Austin could have attested. Serving from 1964 to 1970, Austin likely received the same PSYOPs training as Roger Steffens, who was drafted into the army in 1967 and served two turns of duty as PSYOPs in Saigon. Steffens describes his experience:

> I was twenty-five years of age when I got drafted in 1967. They were taking everyone who wasn't lame or halt. And I enlisted for an extra year because they guaranteed me a job in radio or TV which was my background in civilian life; and instead they sent me to stenography school after basic training, and I was really angry. But as luck would have it, it was at the same small fort in Indianapolis where the Defense Information School was—radio and TV school—and I was able to transfer into that and my first orders were for Ethiopia, to Asmara and Eritrea, to run a helicopter-based radio and TV station, accessible one of two ways: helicopter or safari. And the last week of our school everybody's orders were canceled and we were shipped off *en masse* to Fort Bragg, North Carolina to PSYOPs: Psychological Operations. And we had three intensive weeks of training which began and ended with Nazi films, *Triumph of the Will*, Leni Riefenstahl's—but the four-hour, uncut version: first class, last class. . . . They described it as the most effective propaganda ever made, "Now go to Vietnam and *emulate this*." Those were our orders.[131]

Reviewing a history of postwar British comedy in 1980, Raymond Williams ends with an anecdote.

> When [the Monty Python generation] got to the BBC, from the early 'sixties [*sic*], their chief patron used to refer back to the Berlin revues of the 'twenties [*sic*], and, characteristically, they both accepted this patronage and laughed about the boring historical reference. It seems less boring now, and not only because of what came after it, in Germany, which was usually left out of the flattering retrospect. What is really in question is how we get

through, get out of, a state of disbelief and helplessness which is bound, in all its early stages, to seem comic and edgy: demanding the funny face and the paranoiac prance. Those at least we have had, brilliantly, in a sketch which hopes it will never have to hear the probable punch line.[132]

Williams was writing two years into the Thatcher regime. Today it is hard not to think of the most prominent exponent of the fascist-adjacent America First movement, descending a bronze escalator to announce his candidacy for president of the United States on Bloomsday, 2015. At the time, it was widely received as a put-on or joke.[133] What distinguishes the put-ons of *How Can You Be in Two Places at Once When You're Not Anywhere at All* is that they express the Utopian desire of surrender (yes I will, yes yes), as they simultaneously give voice to disbelief, helplessness, and fear. They also include the punch line.

3 CINEMA / Remediating the Studio System in May 1970

Don't Crush That Dwarf, Hand Me the Pliers (1970)

OVERDUB/EDIT (HOW TO BE IN ONE PLACE TWICE)

Four minutes from the end of the Firesign Theatre's third album, there are thirty seconds of sleight of hand. Passing quickly enough to escape notice on the first few listenings, they constitute the political heart of the entire work, the group's first to span both sides of a long-playing record. For the length of the second side, the narrative has been guided by the conceit of a television's changing channels. We hear pastiches of a morning program, a children's show, advertisements, a soap opera, and, especially, two Hollywood B movies. One is a war film, the other a "lovable, stupid" high school movie; both are meant to be products of the postwar period, the era of anticommunism, the end of the Hollywood studio system's classical phase.

The switching across channels at first seems arbitrary: there are technical difficulties that interrupt *High School Madness*. But after about fifteen minutes, the television begins to switch exclusively, and increasingly quickly, between the two movies. Eventually the sound design abandons the aural cue of the clicking channel. Broadcast on separate stations, the films culminate synchronically in trial scenes. In *Parallel Hell!*, the protagonist, Lieutenant Tirebiter, is court-martialed for failing to give the order to kill. In *High School Madness*, the protagonist, Peorgie, is tried by the newly elected People's Commissioner (alternatively, "dog killer"), who is also his father. Having gone looking for his school, Morse Science High, which has inexplicably disappeared, Peorgie and his friends discover the disassembled building "taken apart, stacked up, and labeled" in the basement of the rival Communist Martyrs High School. Led by Peorgie's father, the authorities find and arrest them; Peorgie stands trial the next day.

The accelerated switching of channels brings the discrete worlds closer and closer, and at last the films briefly interpenetrate—a violation of the unities that is both a narrative and technical feat. The sequence begins unambiguously in the narrative world of *High School Madness*:

DAD: You move when you're told not to move. You take off your shoes to evade the men who are trying to protect you. You sneak into a forbidden sector after curfew, and you are caught with your hands up something they don't belong. My, my, my! Your friends at Commie Martyrs must be mighty proud of you.

PEORGIE: Dad, I don't have any friends at Commie Martyrs.

MUDHEAD: I've never even seen anyone from there!

PEORGIE: You're right, Mudhead. And now there's no room for anybody, 'cause it's all filled up with Morse Science.

When Peorgie and his friend demand to know what has happened to the students of both schools, the scene of Lieutenant Tirebiter's court-martial in *Parallel Hell!* is superimposed on the diegesis of *High School Madness*.

MUDHEAD: Where'd they go?

PEORGIE: Yeah, where did all the kids go?

JUDGE: Well, they're in Korea.

PEORGIE & LT. TIREBITER: *On which side?* . . . [my emphasis]

LT. TIREBITER: I mean, in whose movie?

JUDGE: This is no movie. This is real.

LT. TIREBITER: Which reel?

JUDGE: The last reel of this vintage motion picture, "High School Madness." Lot Number M dash 25. Black and White. 35-millimeter. . . . Now who started at five dollars? . . . Let's go to 25. 25 bid, now 30, now 35 . . .

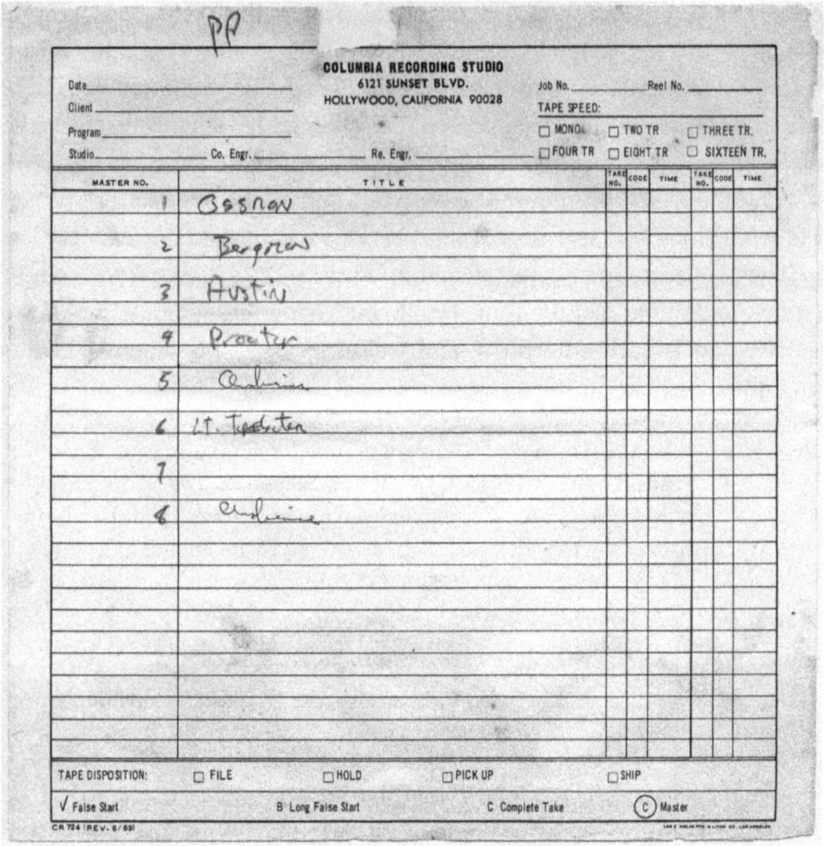

Figure 13. Studio track sheets (reel 31) documenting the recording of the courtroom scenes at the end of *Don't Crush That Dwarf, Hand Me the Pliers* (1970). Firesign Theatre Collection, National Audio-Visual Conservation Center, Library of Congress.

What is undoubtedly an innovation of the writing, as simple quotation shows, is also a technical innovation affecting the record's sound. Up until this point, the rapid intercutting of scenes—the channel surfing effect first used at the end of "How Can You Be in Two Places at Once"—had been produced by the studio technique of the edit, or tape splicing. But to convey the sense of two trial scenes occupying the shared space of a single courtroom required a different technique of tape manipulation—the overdub. David Ossman voices the films' shared line of dialogue, "on which side": he speaks the line as Peorgie in *High School Madness* and, over-dubbed on a parallel track, as Lieutenant Tirebiter in *Parallel Hell!* The diegetic convergence of the films, underscored by Ossman's uncannily doubled voice, required seven of the eight available tracks (some of which likely contained submixes), which were used to place incidental characters and ambient noise in the overall mix, thereby creating the effect of a single, suddenly very crowded, court room (fig. 13).

This sequence was written and recorded early in May 1970, shortly after four students protesting the Nixon administration's announced invasion of Cambodia (which followed revelations in February of a secret war in Laos) were shot and killed by the National Guard, nine others seriously wounded, on the campus of Kent State University in Ohio. In reaction to these killings, "students at 350 colleges went on strike, ROTC buildings were burned on thirty campuses, the National Guard was called out in sixteen states"; at least seventy-five campuses remained closed for the rest of the year.[1] To superimpose a genre film about *war* on a genre film about *school* was to represent allegorically, without naming explicitly, the Kent State killings, as well as the murder of two Black students at Jackson State University eleven days later. Referencing "Which Side Are You On?," the union organizing song made popular by Pete Seeger, Ossman's doubled line suggests students' and soldiers' shared position against the war—with Korea an obvious figure for Vietnam, as it had been in Robert Altman's film from the same year, *M*A*S*H*. (For his part, Richard Nixon had ordered the invasion of Cambodia after obsessively watching another new movie, *Patton*, in the White House screening room.)[2] Firesign was now increasingly aware of their listeners in the US, and in Vietnam, and was writing for both audiences.[3]

By apparent contrast, the device that concludes this sequence—the judge suddenly assuming the role of auctioneer and soliciting bids on the film in which he has been a character—names a second event, exactly contemporaneous with the killing of the student protestors: the Metro-Goldwyn-Mayer auction, which between May 3 and May 21 liquidated forty-six years of the classic Hollywood studio's props and costumes in advance of the sale of its legendary back lot to developers.[4] Although it seems an incongruous and even frivolous association, this abrupt citation of the passing of the old Hollywood was more than a glib device. It was also a marker of Firesign Theatre's intention to remediate the affordances of the classic studio system—from its durable genres to the technical, institutional, and creative procedures of the industry itself—to the cultural opportunities and political exigencies of the moment.

To underscore these two events' simultaneity was a form of "radical juxtaposition," the term Susan Sontag had used to name the enduring inheritance of surrealism, observable by the early 1960s in Alan Kaprow's Happenings, the films of Bruce Conner, and the vogue for Antonin Artaud.[5] It was not only Kent State and the MGM auction that were woven into this record but also the trial of the Chicago Eight ("If you don't answer the question young man, we're going to have to gag you"), the rise of the religious right, the suppression of Black nationalism, the breakup of the Beatles, Richard Nixon, Charles Manson, and (once again) the possibility of a fascist United States. These were all ambient features on a record whose layered narrative complexity, range of reference and implication, and total conceptualization resembled and referenced *literary* models—in particular James Joyce and William S. Burroughs—even as they evoked the formal and sociological conditions of *cinema*: an art form that was, in André Bazin's words, "both popular and industrial," as well as fundamentally collaborative.[6]

TV OR NOT TV

Disquietingly titled *Don't Crush That Dwarf, Hand Me the Pliers*, the Firesign Theatre's third album was their first in which technical, thematic, and narrative elements were comprehensively orchestrated in relation to each other. This accomplishment, I will argue, was owed most of all to their consideration of the cinema as industry and art form. But it also involved their sophisticated understanding of formal practices associated with other media (another way, paradoxically, in which their work resembled cinema). The technique of interruptive tape edits, for instance, clearly evokes the practice of cinematic montage, but it was also directly inspired by the cut-up technique of Burroughs's *The Ticket That Exploded* and by his audiotape experiments, as the group indicated in a radio interview from that same month, May 1970.[7]

On *Don't Crush That Dwarf,* the jarring edits were placed unpredictably in the service of a totalizing, wildly allusive, circular narrative plotted
across a single night: all of which invoked the (cough) "high" modernism
of Joyce's *Finnegans Wake,* and in this way also riffed on their previous
album's direct theft from *Ulysses.* These references across the albums, in
turn, began to craft the conceit of all the Firesign records forming a single
metawork, an idea further developed by devices such as a phone call on
side 1 of *Dwarf* corresponding to one Nick Danger had answered on *How
Can You Be in Two Places at Once.*

For its part, the connection to *Finnegans Wake* was so available to its
audience as to merit mention first in Ed Ward's review for *Rolling Stone*
and the following year in an essay in the journal *College English.*[8] While
these references, and their recognition, say much about the general relation of listening practices to reading practices circa 1970 (as discussed in
chapter 1), it was the record's astonishing sound, abetted by the introduction of Dolby noise reduction technology into the studio, that was most
immediately impressive. Recorded with the increasing comprehension of
Columbia's house engineers, *Don't Crush That Dwarf* elaborated an aural
syntax of narrative space and movement that was new even compared to
the explosive discoveries of *How Can You Be in Two Places at Once.* The
record's audience easily comprehended the effect without having to have
it described in advance. As Dartmouth graduate Robert Christgau (BA
English, 1962) succinctly observed in the *Village Voice,* the group "use[d]
the recording studio at least as brilliantly as any rock group, and there's
really nothing else to say, except that they'd be scary-funny in somebody's
living room, too."[9]

Dwarf is often cited, though not by every fan, as the Firesign Theatre's
masterpiece. It was nominated for a 1971 Hugo Award, despite the fact
that it was neither a book nor a work of science fiction. It was the first
comedy record admitted to the Library of Congress's National Recording
Registry. Christgau awarded Firesign an unprecedented second consecutive A+. Other comedy acts of the period openly borrowed from it, from
the album's depiction of television (Cheech and Chong's debut [1971]
and *Big Bambú* [1972]) to the countercultural courtroom scene (the
Credibility Gap's *Woodschtick* [1971]), albeit with far simpler plotting,
thematics, and sound design. In contrast to those worthy records, *Dwarf*

is possibly Firesign's most complex album, though that's a bit like saying *Highway to Hell* is AC/DC's hardest rocking record, or *Grotesque (After the Gramme)* The Fall's most obstreperous one. Ward's extravagant 6½ column review in *Rolling Stone* (Neil Young's *After the Gold Rush* received a single column in the same issue) valiantly attempts a synopsis before conceding that "to try to summarize the story and its ramifications is almost impossible." His reading of the record's political implications: "time is running out."[10] By way of contrast, Greil Marcus would later name it, obscurely, "the ultimate answer record to *Catcher in the Rye*."[11]

Members of the group have themselves all offered different interpretations, emphasizing by turns media, politics, ontology. That polysemy extends even to the album's title, which has been variously glossed as slang for requesting a roach clip (Ossman), an ethical demand to "fix" the institution of television (Austin), or a tribute to the actors who played *The Wizard of Oz*'s Munchkins and would then fight fascism working in LA's munitions plants during World War II (Proctor).[12] The standard account takes the record as being fundamentally about television, and readers who know the album well will recognize the remarkable similarity of Raymond Williams's description of US television—drawn from the Welsh theorist's 1973 visiting appointment at Stanford—to the structure and events of *Don't Crush That Dwarf*. As Williams observed:

> Advertising has crippled networks and local stations to an extent that is really astonishing, even when the fact is generally known in advance. . . . The habit of interruption has become so strong that most stations also include, within the film you are watching, two or three trailers for films they are showing later. This produces, on occasion, some surrealist effects. Since there is usually no conventional sign for a break to commercial or trailer, a sequence can run from the dinosaur loose in Los Angeles to the deep-voiced woman worrying about keeping her husband with her coffee to the Indians coming over the skyline and a girl in a restaurant in Paris suddenly running from her table to cry. . . . One night there were harrowing pictures of crippled men coming out of the tiger cages in Vietnam and the next pictures in the sequence were of children running in a New England garden to a song about a cereal.[13]

Virtually every one of these images—as well as the "surrealist effects" produced by the vagaries of the broadcast—could be heard three years before Williams's US visit on *Don't Crush That Dwarf, Hand Me the Pliers*.

Largely composed as a pastiche of intercut fragments of broadcast television (news and televangelism in addition to the ones Williams names, as well as fifteen parodic advertisements), the narrative can be initially— but not completely accurately—glossed as representing the experience of a single character who spends the night "watching [him]self on the TV." He begins the day somewhere in his twenties and wakes up the following morning as an old man, before dramatically regressing into a young child as the record fades out. An early draft of this material was titled "A Life in the Day"—another riff on the Beatles but one that could also be an alias for *Ulysses* or *Finnegans Wake*.

One of the record's ambiguities is that this viewer-character is given a name—George Leroy Tirebiter—that he shares, or nearly shares, with several personae seen on the day's television broadcasts, a situation that, if taken seriously, proves to be surprisingly complex, since the Tirebiter avatar is distributed among five or six characters inhabiting at least two discrete levels of reality. A character named Tirebiter is first seen in his apartment (side 1, 8:00), and then on the television *within* the apartment appears both as an elderly game show contestant (side 1, 12:30) and an authoritarian candidate for political office (side 1, 14:55), as well as the fictional lead character in each of the broadcast movies *High School Madness* and *Parallel Hell!* Phil Austin's liner notes for the 1987 Mobile Fidelity Sound Lab reissue explain how these discrete iterations of Tirebiter (on the television and off it) are meant to evoke the "five ages of man," while the group "attempted a scheme whereby the devlish [*sic*] world portrayed in the paintings of [Hieronymus] Bosch was to find its analog in the electronic barrage of television reception . . . a multi-faceted gemlike Hell of transmitted electronic junk controlled by an all too human 'Switcher' able to transform simultaneous transmissions into one sequential program as if he was creating his own life out of the levels of his own internal hell."[14]

These comments about transmissions seeming to coalesce "into one sequential program" uncannily anticipate Williams's influential thesis of televisual "flow" (discussed further in chapter 5).[15] So it might seem perverse to insist that cinema is the most important medium for an album that had also been road tested under the title "The TV Set."[16] But as Williams also observes, the experience of watching television in the early 1970s was to a large extent one of watching old films and made-for-TV movies: "Not so much television, in contemporary California, as telem-ovie."[17] Television scholar Lynn Spigel has similarly argued that the emer-gence of cineastes (film buffs) in the 1950s and 1960s was an effect not only of the rise of art-house cinema but of midnight movies on television; in 1966, Warner Bros. (like MGM, one of the "big five" studios of Hollywood's golden age), had been sold to Seven Arts, which *Variety* mag-azine at the time called "basically a distributor of old pictures to television stations."[18] All of this is to emphasize that the expansion of television did not only or simply threaten the cultural dominance of cinema; it also par-adoxically made cinema and its history newly available. When David Ossman describes *Dwarf* as a "kind of *Sunset Boulevard* [told] in a differ-ent way"—by which he means that George Tirebiter could be understood to be a former actor in the films he watches on television—it is with an awareness both of film history and of the powerful role played by televi-sion for the remediation of Hollywood cinema.[19] Even more than televi-sion, Hollywood cinema, then in a sustained period of institutional crisis, was a formal model for the album and one of its key subjects.

A HEAD OF HIS TIME

To appreciate the meaning of the very high ambitions Firesign Theatre had for their third album, as well as the context in which it was possible to have such ambitions, it is worth looking at the complex and fluid scene of the Los Angeles culture industries in 1969 and 1970—a period in which cinema and popular music were on opposite, but interacting, trajecto-ries—and at the Firesign Theatre's unique position within this scene. Many of the institutional and even technical aspects of these industries suddenly seemed available for negotiation and reconfiguration and could have developed differently. That possibility is emblematized, among other places, in the variability of the phrase "the New Hollywood," which did not in the first instance refer to the new generation of filmmakers like Martin Scorsese or Brian De Palma.

Whereas the sudden institutional dominance of rock music (née rock 'n' roll) was in many ways decisive for the Firesign Theatre, the model of the cinema had been foundational in other ways, not least because of

Firesign's location in Los Angeles. Three years after their first live improvisation of a film festival live on KPFK, the Firesign Theatre were given the opportunity actually to work in the film industry, hired in the summer of 1969 to do a rewrite on a script for a film called *Zachariah*. This experience allowed them to observe directly Hollywood's institutional crisis. Given an office in MGM's Irving Thalberg Building, a nearly empty monument of the old Hollywood, they were able to walk MGM's legendary backlot, encountering the set for *Meet Me in St. Louis* (1944) and "an acre of staircases moldering in the sun," while also finding time to improvise material that would eventually appear on the *Dwarf* album, all under the informal moniker F. Scott Firesign (a joking reference to Fitzgerald's screenwriting and script doctoring day job for MGM in the 1930s).[20]

Firesign's sojourn in the Thalberg Building came just before MGM (on the direction of its new majority owner, casino magnate Kirk Krekorian) would task a former president of CBS television (James T. "the Smiling Cobra" Aubrey) with a dramatic restructuring of the studio. Within two months, the erstwhile mastermind of inane juggernauts *Gilligan's Island* (1964–67) and *The Beverly Hillbillies* (1962–71) announced the backlot's sale to developers and authorized the auction of Dorothy's ruby slippers ($15,000), the Cowardly Lion's skin ($2,400), and thirty thousand other props and costumes from Metro-Goldwyn-Mayer's classic period.[21]

Before writing it into the pages of *Dwarf*, the Firesign Theatre read from the auction catalog on their radio show:

> For sale, four Roman ceremonial axes, one model Russian locomotive (length: three feet, scale: 3/4" to a foot) . . . one whipper's uniform, two Roman swords with scabbards and belt, one cat o' nine tails (63 inches), one slave whip, leather (44 inches), one slave whip, leather (57½ inches), four apothecary jars from *Madame Bovary*, . . . two tub chairs low tufted backs upholstered in green silk, *Cat on a Hot Tin Roof*. One model Russian MiG fighter from *Ice Station Zebra*, one model Martian spaceship (width at stern 24 inches, length 33 inches), one model flying saucer (diameter six inches) from *Forbidden Planet* . . .[22]

And they attended multiple sessions of the auction, acquiring as trophies Jimmy Durante's copy of W. C. Fields's top hat and an inflatable bomb made for the 1960 adaptation of H. G. Wells's *The Time Machine*.

This institutional history helps explain the caustic attitude toward television expressed on *Dwarf*, a position that exists in tension with the record's more ambivalent relation to Hollywood, which sways between cinephilic respect and Adornian critique. And, considering the group's unusual position—in which their pioneering audio work in Columbia Square's Studio B was supplemented with conventional film industry work in the Thalberg Building—it is tempting to suppose that the group at this time was not simply reflecting on the studio system's legacy but also discovering it as a fertile source for their writing. Rather than television, in this view, it could be the recording industry (newly governed by rock) that might inherit and refashion the studio system's cultural position and function, in much the same way late-1930s radio drama had hoped to renew theater as a literary art.[23] It was, after all, music, not television, that had "replaced film as the most important element in youth culture as a whole and especially in the counterculture's core aesthetic and ritual practices," as film scholar David E. James observed in his discussion of a wildly successful film released in March 1970: the *Woodstock* documentary.[24]

The film the Firesign Theatre had been hired to write was further evidence of rock music's new cultural centrality. Originally conceived for Bob Dylan, the Band, and Brigitte Bardot, *Zachariah* took a cue from the landmark film *Easy Rider* (1969), substituting for the latter's nihilist neo-western a camp, pacifist allegory that openly used Hermann Hesse's *Siddhartha* as a source text: the original draft had been penned by writer/director Joe Massot in 1968 after he joined the Beatles at the Maharishi Mahesh Yogi's ashram in India.[25] Such a précis suggests that the process of rock's cultural authorization had not by 1970 completely canceled out its countercultural energy or spirit of play. When released in 1971, *Zachariah* was billed as "the first electric Western" and prominently featured—in lieu of the rootsy Dylan and the Band—the harder and more psychedelic acts Country Joe and the Fish, the James Gang, the New York Rock Ensemble, and White Lightnin' (as well as John Coltrane's drummer Elvin Jones). The film incongruously incorporated features of rock culture within its old west mise-en-scène, opening with a highly entertaining sequence of the James Gang raging through a proto-Stooges jam on the desert floor. Its tagline was "Zachariah—A head of his time" (fig. 14).

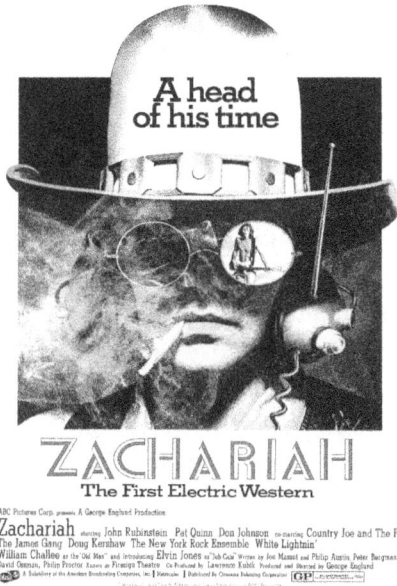

Figure 14. Poster for the first electric Western, *Zachariah*, written by the Firesign Theatre (dir. George Englund, ABC Pictures, 1971).

Notably, *Zachariah*'s burlesque elements did not mask its clear, even principled, opposition to the war in Vietnam and police brutality at home: the film concludes with the two lead characters refusing to duel, choosing instead to live together on something resembling an intentional community. The *New York Times* bemusedly deemed the movie queer propaganda before affirming its visionary *bona fides*: it was "less an aspiration to 'How to Stuff a Wild Bikini' than to 'Zabriskie Point.'"[26] While none of this testifies to the film's enduring greatness, it is still revealing to note how *Zachariah* can be classified among the films that Pauline Kael would soon consecrate as "acid-Western" in her November 1971 review of Alejandro Jodorowsky's *El Topo*: a subgenre that would founder when Sam Peckinpah's MGM feature *Pat Garrett and Billy the Kid* (1973) was released in bowdlerized form, its budget slashed, production rushed, and eighteen minutes cut at the behest of the aforementioned smiling cobra, Aubrey.[27]

A conspicuous, if unsurprising, feature of the acid-Western microgenre is the involvement of rock music throughout: John Lennon and Yoko Ono attended the New York premiere of *El Topo* and persuaded Apple boss Allen Klein to buy its distribution rights; alongside Kris Kristofferson and

Rita Coolidge, Bob Dylan made his first screen appearance in *Pat Garrett and Billy the Kid*, and he wrote the soundtrack; neo-westerns such as *McCabe and Mrs. Miller* and *The Hired Hand* (both 1971) also included soundtracks incongruously identified with the music of the counterculture. None of these films, of course, went all-in with the Rawk in the way that *Zachariah* did (the *Times* review was no doubt taking its cue from Pink Floyd's indelible contribution to Antonioni's *Zabriskie Point*). Taken together, though, the microdot microgenre is remarkable less for the way it exhibits the influence of the newly accredited rock on the emergence of the New Hollywood (heralded by Peckinpah and *Easy Rider*, made doxa by Peter Biskind) than for its testimony to the still-undecided relationship between a not-yet fully incorporated and neutralized rock, on the one hand, and the *old* Hollywood, on the other (as exemplified by the Thalberg Building, the procedures of the old studio system, the durable genre of the western).[28]

Here one need only contrast the goofy left-identified politics and visionary jams of the electric western *Zachariah* with the rehearsed insouciance and relaxed melodies of those revanchist desperados the Eagles, about whom Christgau would very shortly write: "It is no accidental irony that such hard-rock professionals convey their integrated vision of self-possession and pastoral cool by way of a dynamite corporate machine."[29] Indeed, the Eagles' *Desperado* album—for which the band humorlessly posed in sepia tone with pistols and bandolier, in stark contrast with *Zachariah*'s winningly hapless "Cracker gang" as played by Country Joe and the Fish—was released in 1973, the same year that the acid-western microgenre collapsed with Peckinpah's *Pat Garrett*. (Yes, I know Joe Walsh was in the James Gang, see note.)[30]

That same year, a *Time* magazine cover story could report, with still less irony, that "today's pop-rock pantheon is the new Hollywood; its principal gods have filled the void left by the Harlows and Gables."[31] From a belated perspective, *Time*'s pop-rock pantheon/new Hollywood conflation may seem the inexorable, encrusted outcome of the process by which rock had superseded film at the heart of youth's ritual practices. The "dynamite corporate machine" that had managed the apotheosis of the Eagles would in this view be merely a remediation of the system that had once fashioned film celebrity at MGM (known as "the studio of stars"), before moving on to the field of rock for lack of Harlows and Gables. And this star-centered inheritance could be further substantiated with the arrival in Hollywood of that benighted genre, the rock opera, its earliest recorded example dating to 1967, but which reached the screen for the first time in, yes, 1973.[32]

It is worth a further brief digression on this topic since the rock opera has conventionally been understood to be rock's sole foray into long-form narrative, and its emergence on the silver screen would in fact exactly coincide with a moment of crisis and temporary breakup of the Firesign Theatre (discussed in chapter 5). During its foundational years, the primary subject of the rock opera genre was indeed *Time*'s trope of the rock star as demigod (equally conventionally, "royalty"). From the messianism of *Tommy* (album 1969, film 1975), *Jesus Christ Superstar* (1970, 1973), and *The Rise and Fall of Ziggy Stardust and the Spiders from Mars* (1972, 1973) to the bum-kicks dictatorship of *The Wall* (1979, 1982), the rock opera's taste for self-celebrating, self-flagellating, or merely ironic "reflexivity" was at first seen by even the greatest critics as an act of maturity: "the result is a dramatic—even operatic—tension, expressed as noise versus music, between the outside world's destructiveness toward Tommy and his inner peace and growth."[33]

But such first-stage self-awareness was bought at the price of the playfulness and experimentation of an earlier phase of studio recording. Compare the dull, distended narrative and corndog self-referentiality of *Tommy* to the medial playfulness of 1967's *The Who Sell Out*, a proudly Warholian album that celebrated commercial pirate radio with a collection of pop songs and ersatz advertising jingles that were at times stylistically indistinguishable (compare "Mary Anne with the Shaky Hand" with "Odorono") and grotesquely depicted the group's members among giant commodities on the album cover.[34] Here the rocker was no demigod but a diminished shill, a commodity himself. Whereas the Who's double-LP rock operas would briefly signal an apotheosis of accreditation, commensurate with encomia for the Eagles' "musicianship," it is the earlier record that exemplifies an altogether more exploratory, reflexive relationship to its format (the LP) and its medium (the recording studio). This attitude, which understood the LP not only as a *format* but as a *genre* in its own right, originated with *Pet Sounds*, *Sgt. Pepper*, and *The Who Sell Out*, but it extended as broadly as *Freak Out!* and *Absolutely Free*, as well as *We're*

Only in It for the Money, Ogden's Nut Gone Flake, Histoire de Melody Nelson, Arthur, and *Atom Heart Mother*. But the only recorded example that applied such insights to the purpose of long-form narrative, which is also to say to the province of literature and cinema, were the albums of the Firesign Theatre. To say they exhibited a much more complex relation to narrative than did any rock opera is to put it extremely mildly.

STUDIO SYSTEMS C. 1970

As my tendentious salutation of the concept album suggests, what Firesign accomplished in the studio was directly related to the transforming conditions of pop music recording generally. These conditions were determined in a quite specific way in Los Angeles. In June 1967 Christgau wrote that "my choice for our very own Liverpool is Los Angeles," and a year later he reported that LA's music scene was being openly referred to as "The New Hollywood"—some years before that term would be reattributed to a new generation of filmmakers.[35] That same month, a lengthy *Rolling Stone* article, "Los Angeles: A Special Report," documented the way the historic home of commercial cinema had quickly become the new world headquarters of the music industry and the site of rapidly evolving technologies and practices of recording.[36] What film scholar J. D. Connor has studied as "the studios after the studios"—were for a few years *recording studios*.[37]

Although the *Rolling Stone* piece mentioned Frank Zappa as the new occupant of the historic Tom Mix cabin in Laurel Canyon, it otherwise pursued no analogy between Hollywood and rock royalty.[38] Instead, it made available a homology between the old Hollywood studio system and the emerging one for music, systems made up not only of the "talent" and "the best facilities" but also of powerful labels that resembled the old film studios. The overall system included "writers, photographers, artists, critics, producers, marketing consultants, promoters, managers, publicists, messenger services, and at least a hundred other occupational categories" working in concert with "318 record wholesalers and manufacturers listed in the yellow pages."[39] Whereas *Time*'s 1973 cover story would quote Columbia boss Clive Davis asserting "you have the individual emerging again"—a comment that could also have applied to the resurgence of stand-up comedy—the earlier *Rolling Stone* article is noticeably more interested in the overall composition of the studio system and in bands rather than solo artists.[40]

All of this chimes with the understanding of the Firesign Theatre, who were themselves name-checked among the Laurel Canyon scenesters in the *Rolling Stone* article. Within the record industry, they identified not as individuals but as a band, and during their brief Hollywood phase, they identified not as actors but as writers. Although in 1967 Gary Usher had approached Bergman about a record deal with Columbia, Bergman had insisted that the contract be made in the name of the Oz Firesign Theatre. When *Zachariah* completed production in the spring of 1970, the group attempted to suppress their individual names in favor of a collective writing credit: "Written by the Firesign Theatre." To the group's annoyance, the Screen Writers Guild insisted that they use their individual names, but the impetus behind the attempt to take a collective credit testifies not only to the Firesign Theatre's identification as a band but also to their appreciation of a chain of associations between the creative practices of cinema and those of recording. These involved relationships among specific technologies, such as the homologous practices of tape and film editing, but extended meaningfully to collective industrial practice generally.

In *The Recording Angel*, Evan Eisenberg observes that phonography "like film, . . . is a collective art, even if record jackets are less generous with the credits."[41] This point recalls André Bazin's 1957 essay "On the *politique des auteurs*," which famously calls for a "sociological approach to [the] production" of cinematic "genius."[42] Bazin had been responding to critics (aspiring filmmakers) he had mentored at *Cahiers du Cinéma* in the mid-1950s. During these years, the young *Cahiers* critics—Jean-Luc Godard, Francois Truffaut, Éric Rohmer, and others—had polemically elevated American commercial cinema to the status of great literature and claimed for the film's director the status of author. In their view, the greatest Hollywood films should be seen as accomplishments of the heroic auteur, who had succeeded against the odds at flying the restrictive nets of a homogenizing and politically conservative commercial studio system.

Bazin accepted the *Cahiers* critics' promotion of the director as author, and he shared their appreciation for the way the greatest films "[did] not shrink from depicting . . . the contradictions of [American] society" (252).

But he also insisted that Hollywood's greatness owed "not only [to] the greatness of this or that film-maker, but [to] the genius of the system, the richness of its ever-vigorous tradition" (257). Emphasizing the *system's* genius underscored the dangers of an individualist, Romantic conception of authorship—the very category on which, in a different key, the rock opera's Tommy/Jesus/Ziggy/Pink would hang its tormented hat. Though he accepted the conceit of director as author, Bazin advocated a method that could account for the vitality of technological innovations, as well as the literally "corporate" nature of film itself, which drew on a very wide range of collaborators and expertise. Referencing ur-auteur Orson Welles, Bazin insisted that *Citizen Kane* "would never have seen the light of day without the cooperation of some *superb technicians and their just as admirable technical apparatus.* [Cinematographer] Gregg Toland, to mention only one, was more than a little responsible for the final result" (254, my emphasis). Going further, Bazin stressed that although the restrictions of Hollywood's mode of industrial production were real, such confinement was frequently also the "base of operations for creative freedom" (258). Genre itself, he insisted, was an industrially produced, commercial artifact.

With its extended pastiche of two of Hollywood's most durable genres, the high school movie and the war film, as well as its invention of a character, George Leroy Tirebiter, who appears to have been a B-movie actor rather than a star, *Don't Crush That Dwarf* accords thematically with Bazin's anti-Romanticist, sociological account of the studio system. But what is most revealing about *Dwarf*'s engagement with cinema is the way it also corresponds to the technical and collaborative emphases of Bazin's essay. Firesign's use of the tape edit, for instance, by turns inaudible and ostentatious, formed an obvious parallel with film editing, where the assemblage of discrete shots could be used either to enhance the effect of realism (in film, continuity editing) or to emphasize rhetoric or artifice (montage). The analogue provided by the ever-relevant Beatles would be to compare the seamlessly disguised edit that sutured together two discrete takes of "Strawberry Fields Forever" (1967) against the musique concrète techniques overdubbed onto "Tomorrow Never Knows" (1966) or "Revolution 9" (1968).[43] George Martin later described all of these procedures as "making little movies in sound."[44] *Don't Crush That Dwarf* made the association still more literal, employing the edit to move the narrative rapidly from one situation to another, corresponding either to the switching of television channels or to the shifts between levels of narrative reality.

These aurally realized narrative effects were achieved with the collaboration of "superb technicians" like EMI's Geoff Emerick: audio equivalents of *Citizen Kane*'s Toland. For the recording of *Dwarf*, the most important of these technicians were Bill Driml and Jerry Hochman, Columbia engineers whose credits had ranged from the Byrds and Big Brother and the Holding Company to Ray Conniff, The Percy Faith Singers, and Thelonious Monk.[45] (The first sound heard on *Dwarf* is Driml saying "take six is rolling!") These examples, and the practices they invoke, were simultaneously homages to Hollywood conventions and attempts to supersede their enabling example. Whereas *How Can You Be in Two Places at Once* had taken as its main sources the hallmark genres of midcentury radio, *Don't Crush That Dwarf* meditated on the history of cinematic production, from its genres to its institutional structure. (It's no accident that George Tirebiter was awarded "the Academy's coveted Good Sport Award in 1956 for Excellence in Hollywood.")

Employing the television as a frame for these meditations, and invoking, as on *How Can You Be*, the cultural memory of the listening practices appropriate to radio fiction, *Dwarf* also signals the group's attempt to make the recording studio, together with its cognate format the LP, a new medium for long-form narrative. In this way, *Don't Crush That Dwarf* elicited more ontological questions about *medium* itself: both in Marshall McLuhan's sense of remediation ("the 'content' of a medium is always another medium") and in the sociological sense of the term described by Jonathan Sterne.

STEREO AND DOLBY

Sterne disarticulates the doxa of the "technical apparatus" from the activity of the technicians (and other users). As briefly noted in chapter 1, in *The Audible Past*, Sterne defines the term *medium* not as autonomous technology but rather as "the social basis that allows a set of technologies to stand out as a unified thing with clearly defined functions," emphasizing that for this definition, "contingency is the key." Only through processes of repetition and institutionalization can specific social/technical practices "*become media*"—that is, "become associated with technology itself in the minds and practices of users."[46]

To read Sterne's critique of technological determinism after Bazin's critique of his *Cahiers* colleagues' Romantic author-worship is also to recognize the contingency of the "genius-system" of classical Hollywood narrative cinema. And in this way, it brings to mind one of the founding moments of media archaeology—the revisionist "New Film History" of the 1990s, in which scholars like Tom Gunning returned to the early *cinema of attractions* to emphasize the noninevitability of the narrative cinema, which, because of Hollywood, had come to dominate and become associated with all the technologies of film.[47]

For present purposes, this line of thinking also suggests how on or about May 1970—because of the apparent collapse of the classical studio system and the rise of rock within the recording industry—the doxa that determined sound (and other) technologies as *media* were in especially contingent phases and seemed especially available for renegotiation. Technologies available to cinema and to commercial music were adopted unevenly, an unevenness that invoked a longer history in the two industries. Working in this uneven, sometimes aleatory, context was the condition of possibility by which Firesign Theatre aimed to invent a new medium.

Stereo

The histories of multichannel sound in commercial recording and in commercial cinema (for our purposes, *stereo* recording) are particularly instructive because they are surprisingly asymmetrical. In the history of musical recording, stereo mixing was adopted relatively late. Mono records were synonymous with pop music; on radio, the monophonic AM band was similarly the dominant format well into the 1970s. When the first stereo LP was introduced in November 1957, its appearance was accompanied by a publicity campaign—controversially persuading consumers to invest in new equipment—which by the end of the 1960s had succeeded in making stereo the dominant format.[48] As an index of this transition, the Beatles had from their very first album (1963) made mono and stereo mixes of all their recordings. At least through *Sgt. Pepper* (1967), they considered the mono mix to be the primary or authoritative mix and typically only attended the mono mix sessions. In 1968, the White Album was released with mono and stereo mixes that were deliberately different.[49] The Beatles' final studio recording, *Abbey Road* (1969), was released exclusively in stereo.

By contrast, Hollywood cinema had adopted multichannel much earlier, beginning with Walt Disney's *Fantasia* (1940) and with further experiments via the Cinerama and CinemaScope formats in the 1950s. Surprisingly, however, the stereo format was completely abandoned as a tool of narrative cinema during the very years that the music industry was promoting it for music.[50] This remained the case until the late 1970s: even Walter Murch's revolutionary diegetic sound work for such New Hollywood films as *American Graffiti* (1973) and *The Conversation* (1974) would be mixed in mono.[51]

The radical stereophonic diegesis of the Firesign Theatre's records—which we will soon examine in detail—appeared in the contexts both of the beginning of the stereo LP's hegemony, and of the waning years of the functional moratorium on stereo narrative cinema. Their records preceded, and very likely influenced, the celebrated later developments in cinema. But it was a new technology, used by Firesign Theatre for the first time on *Don't Crush That Dwarf*, that is most expressive of the relative openness that allowed the group to imagine the recording studio as a medium for long-form narrative.

Dolby

The original Dolby noise reduction technology (Dolby A) was introduced to recording studios in 1965, though it would not be widely adopted until the turn of the decade (fig. 15). By that time it had become especially desirable because the increasing number of tracks then available in the top commercial studios—for all its innovations *Sgt. Pepper*, like *Waiting for the Electrician*, had only been recorded on four tracks—also introduced the possibility of more tape noise ("hiss") accumulating in the mix. Dolby A

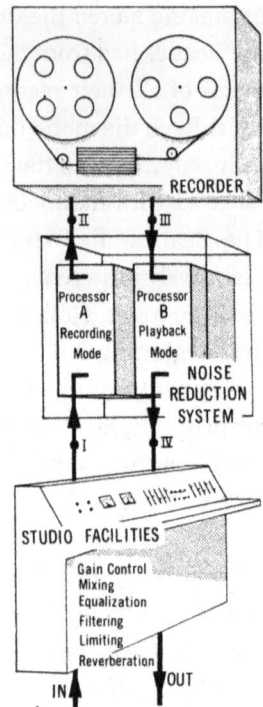

Figure 15. Diagram of Dolby noise-reduction system, produced as an illustration for Ray Dolby's two-part article for *Audio* magazine (June-July 1968).

dramatically reduced this noise. Writing for fellow engineers in 1968, inventor Ray Dolby claimed the new technology provided an increase of ten decibels of dynamic range and was also an efficient means of disguising tape edits.[52] It was especially useful for musical recordings that were compositionally complex but affectively subtle, delicate, or quiet.

When the Elektra Records engineer John Haeny first heard a Dolby-encoded recording, he described "a spiritual awakening"; Elektra boss Jac Holzman called the same event "a technical landmark in pop recording. Incredible clarity, no distracting noise, no veil between you and the music."[53] The recording in question was Judy Collins's *Wildflowers* (1967), an album of earnest, lightly orchestrated pop-folk songs that was without doubt sedulously produced: Joshua Rifkin's delicate string arrangements, for instance, were certainly recorded at a different time and place from Collins's vocal performances. The complexity and fastidiousness of this new institutional arrangement, however (clarity and dynamic range enhanced by the new technology of Dolby), had the paradoxical effect of endowing the singer's voice with stunningly authoritative presence: the "spiritual" authenticity achieved by *Wildflowers* was a historical event accomplished through the means of more superior technological mediation. Whereas Eisenberg stresses concisely that the very noun *record* composes the fiction of a singular live performance, film theorist and rock musician Robert B. Ray has gone so far as to invoke Jacques Derrida's critique of the metaphysics of (Judy Collins's) presence, favorably understanding multitrack recording as instantiating so many modes of writing.[54]

In the field of cinema production, Dolby A was adopted somewhat later: the first film to use Dolby noise reduction on its premixes was Stanley Kubrick's *A Clockwork Orange* (1971).[55] Strikingly, the same currents of thinking—creative embrace of the noise-suppressing technology versus critical demystification—animated practice and criticism, respectively, as Dolby became an increasingly dominant technology within cinema. On one hand, Murch has broadly claimed that it was Dolby that made possible the mid-1970s revolution in cinematic sound: "you could divide film sound in half: there is BD, Before Dolby, and there is AD, After Dolby."[56] Murch is describing here the optical system that became dominant in the late 1970s, commonly called Dolby Stereo, that combined the noise reduction technology with a multichannel (left-center-right-surround) speaker array. Yet even in this more developed context, it was the ability to reduce excess noise (which preceded the doxa of multichannel playback) that attracted the attention of film theorists. In a 1980 essay that cited Murch, theorist Mary Ann Doane critically asserted that "[Dolby] aimed at diminishing the noise of the system, concealing the work of the apparatus, and thus reducing the distance perceived between the object and its representation." In Doane's analysis, the object par excellence was the Hollywood star: "while the desire to bring things closer is certainly exploited in making sound marketable, the qualities of uniqueness and authenticity are not sacrificed. . . . The voice is not detachable from a body which is quite specific—that of the star."[57] On this point and others Doane uncannily anticipates the way Holzman and Haeny will remember Judy Collins's *Wildflowers* recording, though appreciating its effect very differently.

Dwarf *Dolby*

David Ossman has averred that Dolby allowed the Firesign Theatre to achieve "a depth of detail we couldn't have accomplished before."[58] It no doubt aided the superimposition of *Parallel Hell!* onto *High School Madness,* and it increased the dynamic range with which they could work: whereas *How Can You Be in Two Places at Once* is a hectic and loud album, *Dwarf* ranges between soundscapes that are by turns very chaotic and quite intimate; the following year's *I Think We're All Bozos on This Bus* is on the whole rather quiet. Crucially, though—and an index of Firesign's visionary ambition—the core critical presumptions about the new technology, shared by Holzman and Doane alike, do not easily apply to *Don't Crush That Dwarf.* We have already seen the way the Firesign Theatre disavowed any identification with individual stardom (Greil Marcus has rightly remarked that "regardless of photos . . . there was absolutely no way to connect one of the hundreds of voices coming off the albums to a face").[59] But the most apposite point concerns Dolby's relationship to the values associated with fictions of immediacy and presence. Rather than "reducing the distance perceived between the object and its representation" (Holzman's "no veil between you and the music"), the Firesign Theatre understood that Dolby allowed them, contrarily, to depict and engage the illusion of *many more veils.* The soundscapes of *Don't Crush That Dwarf* are as concerned to represent narratively multiple mediations, as they are to encourage the fiction of an unmediated "presence" (something that is, incidentally, true of Murch's early work on *The Conversation* or *THX 1138,* though much less true of blockbusters like *The Godfather* or *Apocalypse Now*).

Ossman has emphasized the way this involved a collaborative reorientation of accepted practices in the studio. The accomplished union engineers at Columbia Square, Driml and Hochman, were experts at producing the fiction of immediacy that had been ecstatically achieved on *Wildflowers.* Ossman describes the challenge this initially presented:

> We had to teach them to be *not* perfect, because everything that we would do would be distorted by some effect. Think for example of the layering of the media when Babe buys a car [on *How Can You Be*]. . . . Each one of those required a different EQ setting. . . . They had to reset each one in order to

sound like a television, a deep FM radio which had to have a deep low end (and then you play a really crappy record on it), and television, and not hi-fi but shortwave, which has its own sound. So each one of those is a degradation of sound, and what the engineers at Columbia were used to doing was making *perfect* sound. So we taught them and worked with them and they loved, finally, doing it.[60]

At a secondary level, Ossman's testimony also indicates the degree to which practices that had been invented and codified for radio drama by the late 1930s and 1940s had been forgotten and required rediscovery three decades later.[61] Yet Firesign's work with Driml and others was very far from a simple act of rediscovery, as indicated by the multiple mediations lovingly described by Ossman. The diegetic effects Firesign achieved were affairs of new and residual media.

An exemplary moment occurs at the beginning of *Dwarf*'s second side. Here, an emergency announcement from the principal of Morse Science High School (on the subject of the school's disappearance) is heard on a car radio, which is a central element in a scene from a film (*High School Madness*) that is being broadcast on a television (as the "Hour of the Wolf Movie"), which we have earlier experienced as one of the sounds emanating in George Tirebiter's apartment. This sequence, which lasts all of seventy seconds, is at least as refined as those Murch will craft for *American Graffiti* three years later.[62] The principal's voice is superimposed on a fight song played by a marching band; together these two elements are EQ'd to produce the effect of an automobile's tinny speakers and mixed together with a third element: a series of bursts of AM radio static; all of this, in turn, is overheard by Peorgie and his friends, who stand around the car and comment on the broadcast ("They're gonna surround us!"). As aurally detailed as the principal's announcement is, it still occupies a discrete space in the mix: comprising three distinct sounds, it is one narrative element in a larger soundscape. The opening of an LP's second side, depicting a television broadcast, which shows a scene from a film, in which students strain to hear a car radio: the effect dramatizes what might be called an *expansio ad absurdum* of McLuhan's famous dictum "the 'content' of any medium is always another medium."[63]

STONER NARRATOLOGY

This dense, multiply mediated technique of intertextuality is undoubtedly why Firesign Theatre records are difficult for some people to hear. A fan letter from a soldier in Vietnam told the group, "You don't get much out of your albums the first few times. . . . We had [*Don't Crush That Dwarf*] about two months before figuring out what it meant."[64] (They had now memorized the album, he affirmed.) On *Dwarf*, narrative movement is frequently accomplished by means of movement among media, and part of the thrill, or the difficulty, of listening to the records is attempting to decide which level of mediation one is meant to be experiencing at any given moment: at a pep rally, from deep in the recesses of the gym, a Morse Science student shouts at the principal: *"What is reality?"* As I hear it (the issue is subject to some interpretation), the story shifts, over the course of the record, among three diegetic levels (fig. 16):

1. An ambiguous space that does not correspond to empirical, embodied experience ("not anywhere at all")

2. A Euclidian social space that includes George Tirebiter's apartment and an evangelical church, the latter being the first "real" location in which listeners can place themselves

3. The highly mobile realm of the television broadcast, which contains *High School Madness* and *Parallel Hell!*

Though the convention of listening collectively and stoned may have made the shifts easier to follow, they are nevertheless meant to be disorienting. And within the third of these narrative realms, the rapid cutting between films, advertisements, and other television programming (accomplished through dozens of edits in the recording studio) is an additional source of disorientation. It was a technique that combined the politically diagnostic juxtapositions of editing associated with avant-garde montage (from Soviet film to Godard's 1968 films) or with Burroughs's tape experiments, while at the same time remaining finally assimilable within a traditional, coherent narrative reality (classical Hollywood style). See, for example, the dramatic edit across the word *change* discussed below; it is a shift between narrative worlds, as well as a reference to the *I Ching* (*Book of Changes*), that could still be rationalized as a simple "change" in television channels.

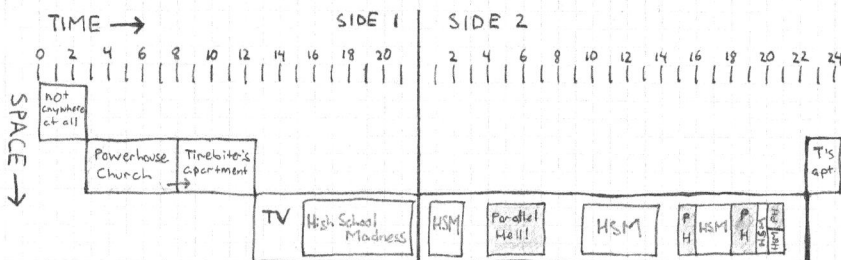

Figure 16. Narrative-diegetic map of *Don't Crush That Dwarf, Hand Me the Pliers*. Author's rendering.

The predominant form of movement on the record is travel across and within media rather than in social space. Any social movement, further-more, is always also an affair of technological mediation. In this way, the album's formal structure emphasizes the general situation of its world, which is a thoroughly mediatized society. Here the group's Adornian ten-dencies appear alongside their cinephilia. We are in a version of Los Angeles, but it is a (semi)counterfactual Los Angeles characterized by quasi-fascist restrictions, administered districts, and lockdowns ("Offer not good after curfew in Sectors R or N"). That the narrative should be so manically peripatetic while its principal character is so constrained sug-gests that *Dwarf* could be read both as a darkly proleptic satire of Williams's thesis about broadcast media as an epiphenomenon of the gen-eral social condition he calls "mobile privatization" *and* as a literalization of the affinities Adorno intuited between the US culture industry and Nazism during the years he lived in Los Angeles.

Overheard in Tirebiter's kitchen: "In other news, final steps was [*sic*] taken in or near Washington to secure the merger of the US Government with TMZ General Corp. This former zinc bushing [click]." The obliquely authoritarian situation of the record further substantiates the uncomfort-able jokes on *How Can You Be* about FDR's "unconditional surrender" after Pearl Harbor and assertions that "the president of the United States *is* named Schicklgruber." Taken together, these create the sense that all the Firesign Theatre albums can be understood to make up a single work that reveals the Adornian underside to the Firesign Theatre's seemingly McLuhanite comedy. On Firesign's *Radio Hour Hour*, one day before the Kent State killings, Philip Proctor read a lengthy quotation from the author of *The Rise and Fall of the Third Reich*, William Shirer: "We might be the first people to go fascist by the democratic vote, and that will be something that not even the Germans or the Italians did. . . . It would be a sort of dictatorship by approval."[65]

(1) *How Can You Be in Two Places at Once* famously opens with the ingratiating, and somewhat annoying, voice of the used-car salesman Ralph Spoilsport. The listener is directly addressed, and although we may at first believe we are hearing it as an advertisement on the television, overall the effect is not meant to be disorienting. *Dwarf*, by contrast, begins with one of the most disconcerting and abrasive openings in the entire history of Columbia Records: 140 seconds that more closely resemble the musique concrète more commonly found on Columbia's Masterworks imprint.

The inaugural moment of studio verité voiced by Driml—"take six is rolling"—is followed by coughs, throat clearing, yawns, inaudible mumbling, and shifting of chairs. A craggy voice, moving into the foreground and from left to right across the stereo field, berates the other indistinct figures and turns on a "chromium switch." We then hear a square-wave tone from a Moog synthesizer (evidently activated by the chromium switch) groan erratically up and down in pitch. This sound is joined by a similarly dyspeptic Hammond B3 organ (the same instrument that had been used to great effect on "Nick Danger"). The Moog and the organ come in and out—the effect is of a power generator warming up—the pitch of the synth vexingly bending up and down as the organ modulates between F major and B flat major chords, and then between F major and C major—"plagal" modulations (typically the "church" cadence, which is a clue) that will take a full eighty seconds before resolving into B flat major (with F in the bass).[66] As this is happening, an audience emerges (via overdub), into what can still only provisionally be called a narrative world, and tentatively applauds. It is not clear what, if anything, is being applauded.

Along with the two keyboards and the audience, a variety of other electrical noises come on and off—identified in the working script as "ELECTRIC VACUUM CLEANER, MIXER, SAW, SHAVER, LAWNMOWER, ETC."—all of which create a distinct sense of unease. According to the same script, "Various mikes at various points in the hall go on (interrupted off and on) amplifying the specific AMBIENT SOUNDS which cut in and out."[67] While these may be instructions to the engineers Driml and Hochman, and some appear not to have made it into the final mix, much of it is audible among the other sounds.

At the one-minute mark, a second speaking voice emerges, sighs with relief ("My iron lung is working again!"), and then disappears. A narrative perspective now begins slowly to emerge. Engineered so as to sound projected through a cheap speaker (the script suggests it was prerecorded to tape cassette), a voice in what can fairly be called aural blackface exhorts: "Well do you know about the gatherin'? I said, do you know about the gatherin'? I say the gatherin' of the Revolutionary Forces! Well, that's gonna be at Reverend Willie's pad—at three o'clock this afternoon—and be on time."

In the foreground, a group of four voices—two present in the room, two mediated as by walkie-talkie—respond to this transmission: "One of those damn kids has got a radio!" / "Bob, this is Mobile Security Patrol One. There seems to be a young-type person in the audience with a Negro radio. Would you check that out please?" The radio is discovered and switched off, the audience applauds again, and now begins to clap in time as the organ finally settles into a 4/4 beat on the right of the stereo field. As the instruments finally coalesce, howling microphone feedback and tapping on a microphone are heard to the left of the sound field, and a speaking voice emerges in what finally, almost 2½ minutes into the record, locates all the aural elements coherently in a single rationalized space: a large evangelical Christian church.

> REV. MOUSE: Is it going to be all right?
> AUDIENCE RESPONSE: It's gonna be all right!
> REV. MOUSE: Ha, ha, ho! You bet, Dear Friends, it is going to be all right. It's going to be all right tonight, here at the Powerhouse Church of the Presumptuous Assumption of the Blinding Light.

The Reverend Deacon E. L. Mouse's voice is treated with reverb and equalization to create the effect of a microphone transmitted through a public address system; the audience response is recorded off-mic to convey the size of the hall and their distance from the dais. Rev. Mouse is panned far to the left; the Hammond organ ("Organ Leroy, at his organ again") is panned hard right; the audience is panned just off center right. We are positioned to observe the service from the side.

Dwarf in this way begins with a radical staging of being "not anywhere at all," a nebulous and unsettling aural situation that eventually resolves into a more conventional diegetic location. Here, the ambiguous elements that will become incorporated into the narrative—handclaps, Hammond organ—compete against elements crucial to defining the initial sound world but will *not* be incorporated into the world of the story: Moog synthesizer, vacuum cleaner, shaver, etc. From one point of view, *all* of these things have been activated by the "chromium switch," but because this is such an ambiguous space, it invites more speculative questions. What has been turned on? The narrative? The most available explanation is metaphorical: the album begins by evoking the title of their first album (*Waiting for the Electrician or Someone Like Him*), which in turn invokes a device that it turns out will run throughout all of their Columbia records: electrical power as a punning figure for social or political power.[68]

However abstract, this accords with the subtext that seems to bind together all of the Columbia records. Power has been turned on, and we have already seen one of its consequences. The "Negro radio," broadcaster of the politically radical "revolutionary forces," has been suppressed and replaced by conservative white evangelism, which now appears to have exclusive access to the airwaves. This can be heard as a reference to the 1969 firing of Roland Young from KSAN (see chapter 2), another attempt by the Firesign Theatre to critically acknowledge and represent the predominant racism of the United States, California in particular. It was also a version of the racial arrogation Joni Mitchell would practice still more notoriously in her blackface performances of the 1970s (which the scholar Eric Lott has both condemned and remarked as the Laurel Canyon scene's only attempt to represent the racism of 1970s LA).[69]

Admitting the ignominy of its performance in the work of four white men, Firesign's aural figure nevertheless anticipates (and in a kindred spirit) Gil Scott-Heron's landmark track "The Revolution Will Not Be Televised" (1971). After Kent State and Jackson State, Philip Proctor read a fake news bulletin on the *Radio Hour Hour*, announcing that Black and white teenagers had broken into the studios of the major TV networks in order to announce the withdrawal of US troops around the world and the replacement of the police by a people's militia ("This is, of course, a joke and is not really happening"). On the following week's show, the group would discuss racial persecution, as Proctor grimly quoted the public statement of LA police chief Ed Davis: "the revolution has begun and we are going to win it."[70]

Don't Crush That Dwarf was the first Firesign Theatre record to be made entirely during Nixon's presidency (hence "Sectors R or N"), and it begins with a pointed metaphorical reference to the conservative backlash against the 1960s, represented by Nixon and then-governor Reagan, and the related rise of evangelical Christianity and televangelism. In 1970 Robert H. Schuller had debuted the massively influential *Hour of Power* television program from Garden Grove in Orange County, the same year Kathryn Kuhlman brought her influential *I Believe in Miracles* program from Pittsburgh to Pasadena (each of these remediating the radio empire built in LA fifty years earlier by the Reverend Aimee Semple McPherson).[71] Nixon, for his part, was especially close to "America's pastor," the media-savvy Billy Graham.[72] Within a year, gonzo journalist Hunter S. Thompson would openly worry about the "Jesus Freak scare" and its role in defeating the political and cultural gains of the previous decade: "What a *horror* to think that all the fine, high action of the Sixties would somehow come down—ten years later—to a gross & mindless echo of Billy Sunday."[73]

Dwarf's complex allegory of political backlash is one of the record's narrative frames, and it has one further constituent element: mandatory consumerism, which the record figures as the literal imperative to eat. At the Powerhouse Church, Rev. Mouse introduces the Pastor Rod Flash, whose sermon is an extended antiphonal riff that grafts the "consumer society" onto a gaudy refiguring of the Eucharist:

FLASH: Godamighty, I'm hungry! Yes! I'm hungry! Safe and sound and hungry!

AUDIENCE: We're hungry!

FLASH: Of course you're hungry! I'm hungry! We're all hungry! So let's eat!

AUDIENCE: Let's eat!

FLASH: And he said the word!

VOICE: What was it?

FLASH: And we ate it! Hot dog! And what was the word?

AUDIENCE: Hot Dog!

FLASH: Hot Dog! Yes, Dear Friends, a mighty Hot Dog is our Lord![74]

Uncannily, just as Hunter Thompson worried about the Jesus freaks, whom he saw to be making inroads into the New Left and counterculture, Ellen Willis would soon observe "an antisixties backlash that parallels (or parodies) the Nixon victory. . . . Turn to *Rolling Stone* and there's [rock critic] Nick Tosches agreeing 'that *eating . . . is more fun than spiritual introspection*'" (my italics).[75]

Developing out of an inchoate screech of lawless electricity into a coherent diegetic space, the first ten minutes of *Don't Crush That Dwarf* synthesize all of these cultural currents in a set of linked figures conjoining a media theory, a racialized figure of political reaction and consumerism. Electricity → white conservative political power → Powerhouse Church of the Presumptuous Assumption → consumption as spirituality.

That the Firesign Theatre mean for eating to be a figure for *media* consumption, as well as commercial consumption, will be made clear in the album's most famous image: Pastor Flash delivering a hot meal through the screen of a television. But we are not quite there yet. To be strict, when the location of the Powerhouse Church first coalesces, it is not clear that it is the site of a broadcast. This will all change very shortly, as the album effects a narrative transition from the space of the Church to George Tirebiter's apartment (where Tirebiter is introduced as a character) and then abruptly moves into the highly mobile realm of the television broadcast.

(II) During the first of these transitions we remain within the second of the three diegetic registers (the diegetic map's Euclidian space). The narrative moves from one social space to another, though it remains in the same level of the real. It links the discrete locations by transferring the first (the Church) into a television that is receiving a broadcast within the second (Tirebiter's apartment). This is accomplished through a combination of overdubs and effects added in the mix (such as equalization and stereo panning). A second map shows how this transition is made both narratologically and technically (fig. 17).

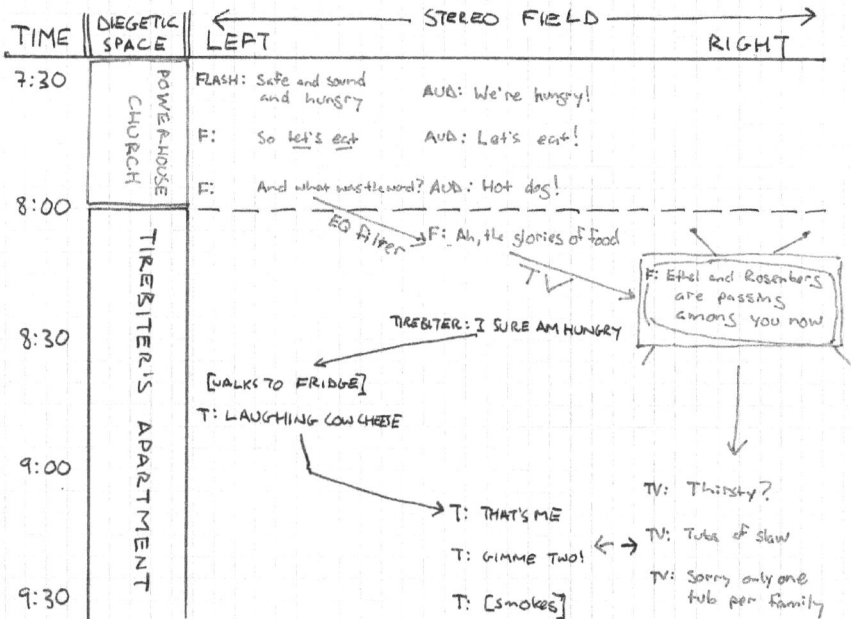

Figure 17. Stereo-diegetic map showing the transition from the Powerhouse Church to George Tirebiter's apartment. Author's rendering.

Picking up just after "A Mighty Hot Dog Is Our Lord" (7:55), the tracks containing the Pastor's voice and the Hammond organ are quickly panned from left to right, and as this happens, they are passed through a filter and turned down in the mix, conveying the effect of now being heard through a cheap speaker: George Tirebiter's television, as the listener soon understands. Like the principal's emergency broadcast, the television is one element in what is evidently a new aural space, one that is close, warm, and muted, in contrast to the reverberant space of the Powerhouse Church. Listeners recognize they have suddenly been placed in a new geography, and at this point (8:10), Tirebiter finally appears. He yawns, sniffs, mutters ("I sure am hungry"), and walks to his refrigerator, his voice and body moving left in the sound field (8:30) as Pastor Flash, suddenly not the primary sonic element, jabbers away on the television on the far right ("Ethel and Rosenberg are passing among you now"). The short sequence dramatizes the Powerhouse Church service as a media broadcast, bringing into relief its oppressively coercive publicity, since it is now heard against the intimacy of Tirebiter's apartment.[76]

One important thematic effect of this is to show how thoroughly public messages permeate private space. Despite Tirebiter's apparent privacy, or isolation, the persistence of Pastor Flash's voice—which the listener is free to follow—shows how the two spaces are not distinct. The Firesign Theatre seize on this for one further narrative and thematic turn of the screw. Turning the channel from Pastor Flash extolling "the glories of food" to find a commercial *advertisement* for food, a hungry George Tirebiter responds to the TV, which serendipitously appears to respond to him (9:10).

ARNIE: Thirsty?
TIREBITER: That's me.
ARNIE: . . . And tubs of slaw
TIREBITER: Gimme two.
ARNIE: Sorry, only one tub per family.

This narrative line reaches an apogee as Tirebiter, having switched back to the Powerhouse Church, now irritably talks back to the Pastor's litany of victuals ("OK, man, you've been talkin' a lot. Hand 'em over!") and, to his shock, receives a hot meal directly through the television screen. Amazed, he immediately starts eating:

FLASH: Say thank you!
GEORGE: Thank you, thank you!
FLASH: Not with your mouth full. I'll talk. You eat. . . . And while you eat, be assured, Dear Friends, that one of the Two Greatest Guys in the Universe, that Great Guy Upstairs, is thanking me, as you are thanking me.

The food is especially welcome, we remember, because Tirebiter's sector is under mandatory curfew. He rapturously eats while engaging in gleeful call-and-response with Pastor Flash:

FLASH: Don't you feel your heart burning?

GEORGE: Oh, I can feel it!

FLASH: . . . Don't you feel the Changes, Dear Friends?

GEORGE: Right on, Daddy-o!

As his affirming cries reach a peak ("Yes, I am ch[anging!]"), the narrative abruptly cuts (a tape edit bisecting the word *change*) to the record's third diegetic register: the realm of the television broadcast that will feature *High School Madness* and *Parallel Hell!*, along with several other programs and a raft of advertisements. Culminating in the near-simultaneous citations of Kent State and the MGM auction, this is where the narrative will remain until an elderly George Tirebiter is awakened, two minutes before the album's end.

EAST COAST / WEST COAST, RADICAL JUXTAPOSITION

Immediately following the release of *Don't Crush That Dwarf*, and for many years after, the Firesign Theatre maintained a contradictory attitude to George Tirebiter. He is the only character to appear on more than one of Firesign's Columbia records, and Ossman continued to write and perform in the aged Tirebiter persona for many years. After Firesign was dropped by Columbia in 1976, Ossman made public appearances as Tirebiter during a performative campaign for Vice President. Yet even before *Dwarf*'s release in July 1970, Bergman was derisively referring to Tirebiter as a version of Faust, and Austin would later affirm: "Faust . . . was to be mirrored in the cowardly George Tirebiter, an all-too-modern figure who sells his soul to the TV in order to buy a coherent view of his own crumbling psyche."[77] Interviewed the following year for *Rolling Stone*, they would go so far as to call him a Berserker ("anybody who's got a gun").[78]

Ambivalence about the "cowardly" Tirebiter was emblematic of a broader uncertainty about the meaning of Firesign's work in general, an uncertainty that seems to have been piqued during their first live tour. Two months before entering the studio for *Dwarf*, they flew east not knowing, as Ossman has recalled, if their first two albums had made any impact outside of California.[79] They began their tour at the end of February at an ecology conference in Washington, DC, and then sold out three consecutive shows at SUNY Stony Brook. They did similarly well at Yale, Bard, Princeton, Columbia, Brown, and Bridgeport University and added a special live broadcast on March 5 from WBAI's Free Music Store on East 62nd Street in New York City. An ad in the Columbia student newspaper, twice quoting their most recent album (including their riff on SDS's "bring the war home"), showed the way the group's language had already become a kind of lingua franca among students at universities most associated with the antiwar movement. Austin remembers "walking on stage at Columbia University, through barricades and people raising their fists at us."[80]

Popular as they were, the Firesign Theatre was also met with some skepticism, particularly on the question of politics. Reviewing the WBAI and Columbia performances, the *Village Voice*'s Susan L. Pansey acknowledged Firesign's aesthetic power but was exacting on the issue of the group's commitment: "The demands are real, the confrontation heavy, but sometimes the whole premise wobbles dangerously near the abstract meaningless, too caught up in its own universality. . . . Where 'mystical communication' leaves off and complete mystification takes up can be a very subtle place. . . . Getting to be in two places at once may be a good creative ideal, but I for one don't happen to think it's possible and/or desirable when those two places are revolutionary and counter-revolutionary, and once involved and standing aside laughing."[81]

Annoyed by the group's evasiveness in her interviews, Pansey did not mention an eight-minute piece on Native American genocide that involved harrowing eye-witness accounts of the Trail of Tears, which were read directly from oral histories and for no comic effect (a variation on the November 1967 performance discussed in chapter 1). This was followed by a short caustic satire of white enjoyment and then an antiphonal reading of David Antin's incantatory "History" (1968), a poem that begins with a long list of international political assassinations; enumerates instances of shipwrecks, arson, and sabotage; and ends with the names of six children, each of whom had been abducted and never found.[82]

In general, the East Coast institutions (*Village Voice*, WBAI) that were cognates to those most supportive of the Firesign Theatre on the West Coast

(*LA Free Press*, KPFK) were also the ones most likely to insist that the group assert a political position. This revealed a fundamental difference of political style and even consciousness (the group, along with Annalee Austin and Tiny Ossman, were discussing feminism for some weeks afterward on KPPC).[83] And there were concurrent events that brought the matter to the fore. The day after the Free Music Store show, three radical activists from the Weatherman group had blown up themselves and a West Village townhouse, when a bomb they were making exploded prematurely; three days after that, SNCC activists Ralph Featherstone and William "Che" Payne were killed en route to the Maryland trial of H. Rapp Brown when a bomb exploded beneath their car.[84] The day after that, with political death and violence in the air, the Firesign Theatre were Paul Gorman's guests on Pacifica's WBAI, where Ossman had hosted poetry programs a decade earlier. Gorman had helped organize Firesign's tour and was clearly a fan, but he had also worked with Featherstone in Washington, DC. Before going on the air, he had just spoken with his friend Paul Cowan, who had worked with Featherstone as a Freedom Summer volunteer:

GORMAN: He suggested that I ask you guys . . . exactly how you feel about politics, and how you react to criticisms that have come from people about being apolitical—how you guys relate to events that happen, because I went through some funny changes because that part of me that is relating to how you guys see the world is different from that part of me which relates to immediate struggle and events like [Featherstone's death].

PROCTOR: Why?

GORMAN: Why is it different?

PROCTOR: Why is it different?

GORMAN: Because in order for you guys, I think, to have the comic vision that you have, I think it requires a stance of independence from events.

PROCTOR: Yeah.

GORMAN: A stance from trying to relate to those events as "something that happens," as even karma.

PROCTOR: Yeah.

GORMAN: That beneath your work is a certain level of acceptance and detachment that is very different . . .

PROCTOR: That's one *aspect* of it.

GORMAN: ... from, say, Paul Cowan whose relationship to events like this ...

PROCTOR: All right let me talk. The major difference is—for one thing we express our philosophies through our humor so this is nonsense, to really sit and rap about this, for us and we don't like to do it.

GORMAN: But it's nonsense that people want to hear.

PROCTOR: Well, of course, but why do people want to hear it? Because they *expect* to hear it on a political platform as opposed to, you know, the platform of your own head, of the visions through which these things come: parables. We speak in parables. ...

AUSTIN: We defy anybody to bring us down, the fascists or anybody else. The Firesign Theatre is not "clown" per se. We are not crying behind our masks. There is neither a serious side to us nor a happy side to us, there is only this continual discussion that we go through within ourselves about all the things that you are asking, and the expression about them on our records comedically is generally the way we see the world.

GORMAN: And I happen to think that's a political view. But the point I'm trying to make is that I really *do* think that there's a different posture that you guys *should* have, have to have, and *do* have as artists ... with relation to political struggle. I think you relate to it, but I think a lot of people get a flash of a certain kind of detachment.[85]

Two problems are at issue for Gorman and for Pansey. On one hand is the question of whether the autonomy ("independence") and difficulty ("the demands are real") are inherently counter-revolutionary. Proctor asks if the parable can be political: would it be more or less politically effective to have Kent State as a mediated reference than as one that is named outright? The second question is whether laughter can be politically progressive. It seems likely that the more systematic elaboration of the authoritarian Los Angeles in *Dwarf*—as well as Tirebiter's ambiguous position within it—was designed to express the problem and work out a response.

As rattled as they were by the more intense politicization of the East Coast—Austin seemed particularly insecure when confronted with feminism—they were equally spooked by the reception of some of their devoted fans. In a September 1970 interview with the Boston paper *Fusion*, the group recalled their surprise in meeting fans who had memorized their albums, but they eventually discussed another form of reception:

OSSMAN: Do you realize how serious your work is, how much it's changed my life?

PROCTOR: When I dropped that STP and listened to your album, you really scared the shit out of me. Well, what do you *think* about that?

OSSMAN: Yeah, right. I was asked by somebody, what did I think about the *effect* that we had on people's minds?

BERGMAN: People *believe* that you can have an effect on people's minds, which is bullshit. Like Charlie Manson could possibly *effect* anyone to kill someone. That's bullshit. That's just not true. There's *nothing* you could put on a record player that is going to take your mind over.[86]

The unspoken referent in this exchange—which the group at first seems reluctant to air—is the Firesign Theatre's long identification with the Beatles and the way this identification had been chillingly reframed by the Manson murders of August 1969. The murders had touched everyone in the Firesign circle, especially Proctor, whose wife, Sheilah Wells, had been a close friend of Sharon Tate. By the end of 1969, the radical wing of SDS was understood to be claiming Manson as a hero.[87] When the Manson case came to trial in the fall of 1970, defense lawyer Ronald Hughes wore a double-breasted suit still bearing the tag "MGM Auction—Spencer Tracy."[88] It was an ominous time for art that had so openly solicited hermeneutics.

By the time Firesign Theatre were interviewed for a three-page spread in *Rolling Stone* the following year, they had a somewhat more convincing set of answers to the political question (and were working on a record that, for better or worse, would dramatize the "breaking" of the president of the United States). Asked about the influence of Lenny Bruce, whose *What I Was Arrested For* had just been posthumously released, they answered: "Bruce was a traditional American comedian, that is, a man in pain. We're not part of that tradition. . . . We have more in common with surrealists than satirists. W. C. Fields, the Marx Brothers, Marcel Duchamp, Marcel Marceau, the old Dadaists."[89]

I remarked in chapter 1 the long history of the "alliance between modernist high culture and popular song," the genealogy Bernard Gendron traces from the late 1960s rock album backward through bebop to the moment of Dada (and its "effective fusion of cutting-edge high culture and vaudeville-style entertainment").[90] But the most illuminating citation in the *Rolling Stone* interview—one that affords a metatheory of *Dwarf*, containing and contextualizing its interest in cinema—is that of Dada's inheritors, the surrealists.[91] Here the Firesign Theatre are not specifically invoking Breton, Dalí, and Magritte so much as "a mode of sensibility which cuts across all the arts in the 20th century," as Susan Sontag identified it in her 1962 essay on Allan Kaprow's Happenings. "The Surrealist tradition in all these arts," Sontag argued, "is united by the idea of destroying conventional meanings, and creating new meanings or counter-meanings through radical juxtaposition (the 'collage principle')."[92] Rhyming with cinematic montage, and abetted by the techniques of tape editing, the practices of radical juxtaposition define the aesthetic and thematic character of *Don't Crush That Dwarf* and the Firesign Theatre, circa May 1970. It is a way of naming what annoys the *Village Voice* reviewer ("revolutionary and counter-revolutionary"; Indian genocide and all the puerile jokes), as well as the confluence of Kent State and the MGM auction. It can even be found in the names of the characters in *High School Madness*, Peorgie and Mudhead. Travestying the comic book characters Archie and Jughead, the name *Peorgie* is derived from George (Georgie) Tirebiter, which had been the name of the mascot dog of the USC football team. *Mudhead*, however, or *Koyemsi*, is the name of a sacred clown of the Hopi Indians, as Firesign well knew, a reference that in itself embodies a form of radical juxtaposition: "Like simpletons or the crazed they spoke nonsense one instant and wise words and prophecies the next and as such they became the attendants and interpreters of the kachinas."[93]

Sontag goes on to specify that surrealism, so understood, has been used both for the purposes of "inane, childish" wit and for tendentious social satire. But Sontag also discerns another, deeper vocation for the surrealist tradition: "It can be conceived more seriously, therapeutically—for the purpose of reeducating the senses (in art) or the character (in psychoanalysis). And finally, it can be made to serve the purposes of terror." What is finally most interesting is the fact that, for Sontag, it is these deeper functions of surrealism that bring it closest to comedy:

This art form which is designed to stir the modern audience from its cozy emotional anesthesia operates with images of anesthetized persons, acting in a kind of slow-motion disjunction with each other, and gives us an image of action characterized above all by ceremoniousness and ineffectuality. At this point *the Surrealist arts of terror link up with the deepest meaning of comedy*: the assertion of invulnerability. In the heart of comedy, there is emotional anesthesia. What permits us to laugh at painful and grotesque events is that we observe that the people to whom these events happen *are really underreacting....* This is the secret of such different examples of comedy as Aristophanes' *The Clouds, Gulliver's Travels*, Tex Avery cartoons, *Candide, Kind Hearts and Coronets*, the films of Buster Keaton, *Ubu Roi*, the Goon Show.[94]

It is conceivable that one or more of the Firesign Theatre had read Sontag's essay (it had been collected in the influential 1966 *Against Interpretation*). But whether or not they had, it is striking to observe the explanatory power of Sontag's theory for *Dwarf*, which extends beyond the formal device of "radical juxtaposition."

Several of Sontag's examples are in fact involved in the record, from *Candide* ("the 38th of Cunegonde") to Tex Avery ("A Life in the Day" and "The TV Set" each included aural versions of Sylvester and Tweety cartoons) to the Goons and Buster Keaton. When the aged Tirebiter awakes at the end of the record, his answering service reads him messages from Keaton, Mack Sennett, Laurel and Hardy, W. C. Fields, Hal Roach, Ben Turpin, Stepin Fetchit, Harpo Marx, Harold Lloyd, and Charlie Chaplin: great comedians from the early decades of Hollywood film. Like them, George Tirebiter is without doubt an "anesthetized person" who is "underreacting," given what the attentive listener understands about the general situation of the world of the record. This is true whether one understands him as protagonist or as the cowardly, all-too-human Faust.

In the spring of 1970, Gerald V. Casale and Mark Mothersbaugh—who would go on to satirize the emotional anesthesia of groupthink with their band Devo—were hippies and students at Kent State University. On May 4, they were among the students protesting the invasion of Cambodia. Casale, an active member of SDS, was a friend of two of the students killed that day—Allison Krause and Jeffrey Miller. "For me it was the turning point," Casale remembered. "Suddenly I saw it all clearly: All these kids with their idealism, it was very naive. After Kent, it seemed like you could either join a guerrilla group like the Weather Underground, actually try assassinating some of these evil people—the way *they* had murdered anybody in the sixties who'd tried to make a difference—or you could just make some kind of wacked-out creative Dada response. Which is what Devo did."[95]

Or, as another forebear, the comedian Lord Buckley, said of the atomic bomb (and as both Devo and Firesign would have understood in the context of Vietnam): "I say it is the duty of the humor of any given nation in time of high crisis to attack the catastrophe that faces it in such a manner as to cause the people to laugh at it in such a way that they do not die before they get killed."[96]

4 ARTIFICIAL INTELLIGENCE / Up Against the Wall of Science

I Think We're All Bozos on This Bus (1971)

Don't Crush That Dwarf, Hand Me the Pliers is loosely focused through a character, George Tirebiter, whose multiple personae apparently express a world thoroughly determined by media. As a young man (men) he seems to have been characters in the movie *High School Madness* and in the Korean War film *Parallel Hell!*, both of which are now being broadcast on television. Near the end of the album an elderly Tirebiter tells his answering service, "I've been up all night, watching myself on the TV." Earlier, we heard a younger Tirebiter watching a geriatric contestant (also named George Tirebiter) on a game show; when asked "How does an old man like you stay alive?," he answers, "Well sir, I try to get up every morning and watch television all day." And when George at last leaves his house or apartment, having learned from his answering service that he's missed calls from the great comedians of early cinema, he audibly transforms into a child and chases an ice cream truck as the record fades out . . . "wait for me, hey mister, I got a nickel, wait for me . . ."

If this sentimental (but arresting) image suggests a nostalgia for a world before television, Tirebiter's multiplicity suggests that the Firesign

Theatre was still more concerned with representing the way all experience had by 1970 become thoroughly mediated by technologies of electronic communication. This is why the album invited psychological readings such as the one Phil Austin ventured about "the cowardly George Tirebiter[,] . . . an all too human 'Switcher' able to transform simultaneous transmissions into one sequential program as if he was creating his own life out of the levels of his own internal hell."[1] The same reading rationalized the record's key representation of the Kent State killings—the aural conflation of *Parallel Hell!* and *High School Madness*—as the effect of Tirebiter's having fallen asleep and dreamed the two movies together. In this interpretation, the most important point about television is what it does to your head.

But *Dwarf* can also be understood, less psychologically, as exploring the phenomenon Raymond Williams would name—in a book that itself includes a famous account of narcoleptic television watching—mobile privatization. *Television: Technology and Cultural Form* (1974) begins by considering the twentieth-century phenomenon of broadcasting in the context of "two apparently paradoxical yet deeply connected tendencies of modern industrial urban living: on the one hand mobility, on the other hand the more apparently self sufficient family home."[2] The latter tendency, privatization that could tend toward alienation (as with Tirebiter's apartment), created an "imperative need for new kinds of contact" and "a new kind of 'communication': news from 'outside,' from otherwise inaccessible sources."[3] This new kind of communication was *broadcasting*, instantiated first by radio in the 1920s and later by television, a genealogy the Firesign Theatre would have recognized. *Dwarf* is a vivid depiction of this very sense of media mobility relieving the privatized isolation of domestic space.

Dwarf is very obviously concerned about the extreme forms mobile privatization might take, concerns emphasized by the album's setting in a recognizable though an unnamed near-future Los Angeles, a dystopian state in which Tirebiter apparently never, or rarely, leaves his home, subject to curfew and located ambiguously in Sector R or N; where he has also apparently once run for office as a law-and-order candidate (fronting the "Tirebiter for Political Solutions Committee") but can now be heard haplessly rummaging in his fridge past the mescaline and Laughing Cow cheese, a televangelist keeping him company in the background. The main point here is that it is *not* all happening in his head. And anyway, having the television on would also be a good way to avoid being monitored by electronic surveillance—if Tirebiter had another person to talk to.

The Firesign Theatre's next album would take place in this same parallel semitotalitarian present or near future, and it would indeed represent the phenomenon of tape surveillance, which had become much more than a hypothetical issue in the months following Kent State. Titled *I Think We're All Bozos on This Bus*, the new album made the link explicit by opening with a fading echo of the ice cream truck bells that were the last sound heard on *Dwarf.* But rather than positing a domestic space saturated with transmissions from outside, *Bozos* takes place—save for a few minutes on each end that are set in a *Godot*-like wasteland—in the highly administered space of an amusement park or exposition called The Future Fair. Not a private space riven by publicity, as in *Dwarf,* it is a public space (or hyperreal simulation of one) that is ordered, regulated, and surveilled by media technologies, staffed by engaging holograms and robots, and ultimately determined by private interests. It is a world where "technical stimulation" is a synonym for "government-inflicted simulation," but the government is an exhibit at the Fair and not the other way around.

Because the album's primary media archaeological object is an early and influential example of artificial intelligence, Joseph Weizenbaum's foundational mid-1960s chatbot ELIZA, it is impossible today not to hear the album as a premonitory allegory of the future social power of the tech industries. *I Think We're All Bozos on This Bus* was released the year after Kent State and was concerned with the fate of the student movement as the Vietnam War seemed increasingly impervious to protest. But 1971 was also the year the name "Silicon Valley" was coined (in the trade magazine *Electronic News*); and it was the year a much larger complement to California's original Disneyland opened in Orlando, Florida ("the model of an urban agglomerate of the future," as Umberto Eco would name it a couple of years later).[4] The Firesign Theatre addressed these events together allegorically and critically in their technologized art. Such criticality, meanwhile, would be notoriously absent at an event actually titled "Art and Technology," a monumental exhibition that opened in May 1971 at the Los Angeles County Museum of Art, as Firesign was writing and recording *Bozos.* "Art and Technology" featured collaborations brokered by LACMA between prominent artists and industrial corporations, many of which were materially sustaining the war in Vietnam.

I Think We're All Bozos on This Bus literally began where *Don't Crush That Dwarf* had ended, but the Firesign Theatre were determined to make a record that was not an obvious retread of its predecessor, despite its huge success, and they made a series of programmatic decisions designed to give the new album a distinct feel and identity. As opposed to the diverse references that constitute *Dwarf*—the MGM auction, Kent State, napalm(olive), old movies, televangelism, game shows, and so on—*Bozos* has two clear principal sources: Weizenbaum's ELIZA program (which Phil Proctor had experienced at an LA work fair in 1970) and the Chicago Century of Progress World's Fair of 1933 (whose *Official Guide Book* they took to the studio).

Along with this more restricted set of sources, which accreted related references like Disneyland, Firesign also chose to use a simpler set of formal procedures when they commenced recording at Columbia Square. Turning away from the fragmented and stratified narrative of *Dwarf*, which had been achieved through dozens of tape edits and splices, they decided that the story of *Bozos* should be linear and recorded on spools of two-inch tape that they agreed not to cut. They would allow themselves the luxury of layering sounds through overdubbing, but each recording session would begin by rolling back the tape and starting where they had left off.[5]

These decisions resulted in a more straightforward plot organized around a more traditional protagonist. *Dwarf* and *How Can You Be in Two Places at Once* could each have been seen to express aversion, if not hostility, to conventional characterization. The naive protagonist of *How Can You Be in Two Places at Once* is drafted, changes race, and then totally disappears ten minutes before the end of side 1; Firesign then created the "cowardly" and multiform George Tirebiter, about whom they would express ambivalence and sometimes disdain. At times, it is hard not to hear these character experiments as attempts to expunge or atone for their privilege ("Peorgie, you're a white man, you've got to help us"). But on *Bozos*, Phil Proctor's "Clem" would be less ambiguously drawn as a protagonist, a disgruntled employee (we eventually learn) who destroys the Future Fair by sabotaging its computers. He is also, with the possible exception of Phil Austin's traveler on *Electrician*, the first Firesign character with no parodic vocal markings or accent. Deliberately weakly defined, Clem is blandly "universal," which is to say implicitly white rather than a satire of whiteness.

These decisions made the album available to be read by the 1980s as a proleptic romanticization of hacker culture, and as we will see, hidden references to *Bozos* have been baked into the Apple/Macintosh infrastructure for decades. This was not the dominant reading of the album in the early 1970s, however, and it is less likely to be our reading today. Neither interested in or consoled by Clem's geeked-out ingenuity, *Bozos'* contemporary audience was instead more drawn to the technological-social surveillance-amusement culture of the Future Fair. The Fair's culminating exhibit was an audio-animatronic president who may also be the real president, an absurd but chilling image that would anticipate Joseph Weizenbaum's later apostasy with respect to his own invention and to the concept of artificial intelligence generally. *Creem*'s Dave Marsh (who had attended Weizenbaum's alma mater, Wayne State University in Detroit) agreed: "This is Autonomous Technology speaking, and it doesn't answer questions, except on its own terms."[6] For Weizenbaum, as for the first listeners of *Bozos*, the broader issue would have been the way the confusion of human intelligence with the performance of machines—what today is still called "the ELIZA effect"—was coextensive with a world in which what would later be called surveillance capitalism could be naturalized as a form of entertainment.

INTO THE SEVENTIES, FOR REAL (SLIGHT RETURN)

In November 1970, the Firesign Theatre performed a week of standing-room-only shows at the Ash Grove in West Hollywood.[7] It had been three months since the release of *Don't Crush That Dwarf*, and they had also revived their live radio program. Now called *Dear Friends*, the show had brought them back to KPFK under an arrangement that allowed them to syndicate the best shows, via transcription disc, not only to the other Pacifica affiliates but to dozens of commercial underground and college stations across the US.[8] Following them from the *Radio Hour Hour* on KPPC was the Live Earl Jive (Vaughn Filkins), who engineered the shows and became increasingly involved as an on-air collaborator, responding to and inciting the group's verbal riffs by mixing in sound effects, tape echo, and a very wide range of music.[9] Excerpted as a double LP set in 1972, the September 1970–February 1971 *Dear Friends* shows are often regarded as the peak of Firesign's radio career and contain many of the ideas that would be developed on *I Think We're All Bozos on This Bus*. The group began tracking the album in April, once again beginning without a complete script, and worked steadily until final mixdown in June. *Bozos* appeared exactly a year after *Dwarf* in August 1971. Though they didn't know it, it would be the last album they would track at Columbia Square.

The upward arc of this narrative notwithstanding, these were also months of turmoil as well as grief for the group. Peter Bergman's girlfriend in 1971 was a Vietnam War correspondent named Linda Sugarman. Traumatized by the atrocities she had seen, that spring she took an intentional overdose and died in Bergman's arms.[10] Linda Sugarman's death had followed the equally sudden death of the Firesign Theatre's friend and collaborator Jere Brian (a.k.a. Franklin Delano Wacco). Firesign had broken their association with manager James William Guercio in February and were going it alone (hence the new arrangement at KPFK). Since the break with Guercio, they had also been working on a film project, first titled *Eat* then retitled *The Big Suitcase of 1969*, with Brian raising funds as producer.[11] For several years, Jere Brian had lived on a ramshackle property in the Hollywood Hills near Mount Cahuenga known as The Farm. Bergman, Phil, and Annalee Austin had all lived at the Farm around 1969, its other denizens including the Modern Folk Quartet's Cyrus Faryar and Henry Diltz, Renais Faryar, John and Catherine Sebastian, "Tie Dye Annie" Thomas, the producer John Simon, photographer Jerry de Wilde, Michael Gwynne, Anton Greene, and the hippie Volkswagen dealer Jack Poet.[12] Brian/Wacco was still living at the Farm when, two days before the Firesign Theatre began tracking *Bozos*, he was caught by an undertow swimming in the ocean and drowned. The plans for the Firesign movie were shelved.

Early 1971 was also a time of profound transition in the record industry. With the Beatles having officially broken up in December, the Rolling Stones chose April Fool's Day to announce a massive new contract with Atlantic Records, laying a marker for the normalization of rock superstardom and the metastasizing of the record industry.[13] *Sticky Fingers*, their first Atlantic record and the first to feature their iconic lips-and-tongue logo, appeared in May. By the end of the year the Stones (by then living as tax exiles in France) would be finishing *Exile on Main Street* at Sunset Sound in Hollywood, three miles up from Tower Records, "the world's largest record store," which had opened that spring. What had begun almost two decades earlier as "rock 'n' roll," and then culturally accredited as "rock," was now becoming unambiguously corporatized.

Though these developments would eventually result in their marginalization, the Firesign Theatre remained singularly well positioned in the popular music field when *Bozos* appeared that summer. They had a powerful label that advertised them alongside rock acts, airplay on an FM radio that had not yet been fully restructured as the industrialized and restrictive AOR (see chapter 2), and increasingly expansive critical encomia from the rock and underground press, which understood their records as integrally bound to the world of music. On the heels of a three-page profile in *Rolling Stone* in September, *Bozos* reached number fifty on the *Billboard* chart, twelve spots below *Sticky Fingers*.[14]

Firesign's most discerning reviewers had often heard the records more as frightening social critique than as light comic relief. With *Bozos*, this became the consensus understanding, though for some early supporters, it would be grounds for critique: Robert Christgau wrote that with this album, the Firesign Theatre had "abandon[ed] humor for sci-fi middle seriousness" and gave them (ouch) a B–; Greil Marcus, responding to the plot's linearity, called it "ambitious, overly rational," and "spooky, but a little too obvious."[15] These complaints were not groundless, but they were far outweighed by breathlessly appreciative reviews, which were very large in number and by now international. In London, where the record was only available as an import, *Bozos* merited the lead review both in the liberationist paper *Oz* and in the *International Times* (*IT*). The latter read the album alongside the philosopher Henri Bergson, averred the Firesign Theatre's "very revolutionary force," said Monty Python looked uptight in comparison, and insisted on the "total concentration" *Bozos* demanded: "Its power is astonishing: a friend played it over Christmas and found it impossible to talk to anyone afterwards."[16] *Rolling Stone* named *Bozos* "a modern *Hard Times*"; the radical Chicago paper *The Seed* would call it "very, very deep" and "as scary as a bad LSD trip."[17] Back in Detroit, it was Marsh who went the furthest:

> It's not funny, as I was informed even before I'd heard it, it's terrifying. It is—to make a long story short—the future. . . . This is the most direct and forceful statement that Bergman, Ossman, Proctor and Austin have laid on us, yet it's still so multi-level it boggles the mind. . . . The dire tenor of the situation is a whole lot clearer than it was before. . . . You could chalk it up to fantasy, but you'd be making a mistake. . . . There is no question in my mind that the Firesign Theatre have a better idea than I do how urgent the situation is, how far the rot has progressed, to what degree their statements are not futuristic but right NOW.[18]

Xgau's and Marcus's gimlet-eyed dissent notwithstanding, these astonished salutations of a dystopian futuristic comedy album that was apparently not funny can be understood as evidence of a record that met its moment, though not in an obvious way. Even as 1971 was seeing the corporatization of rock music, it was also, and relatedly, a period of uncertainty for the counterculture and the antiwar movements, and this uncertainty resonated particularly on the question of media's relation to politics. As we have seen, key actors in the movement had extolled the availability and seemingly easy manipulability of mass media. The Yippies had organized the protests at the 1968 DNC largely by calling in to New York's WBAI and would later phone in live daily updates from the Chicago Eight trial.[19] Television and print coverage were each important for the cultural dissemination of the Yippies' satirical and often silly put-ons, but when Jerry Rubin said that "you can't be a revolutionary without a television set," he was thinking not only about Abbie Hoffman levitating the Pentagon but about the Yippies' on-air clashes with talk show hosts and the televised chaos in Chicago.[20] In Ellen Willis's words, the Yippies' provocations "got the Chicago police and the news media to cooperate in bringing revolutionary theater to millions."[21]

Three hundred Yippies invaded Disneyland in August 1970, raising the NLF flag over Tom Sawyer's Island and chanting "Ho, Ho, Ho Chi Minh."[22] But these sensational stunts—which the Firesign Theatre had explored with a more reflexive eye—had begun to decisively wane by 1971. John Lennon and Yoko Ono had staged performative happenings for peace and against stereotyping in the spring of 1969 (the Amsterdam and Montreal Bed-Ins, Vienna's "Bagism" event). Their affect was much more soberly militant in December 1971, when they headlined the John Sinclair Freedom Rally in Ann Arbor, performing new songs about the Attica prison uprising and about the prison sentence of the White Panther leader: "We came here not only to help John and spotlight what's going on, but to say to all of you: Apathy isn't it, and that we can do *something*. OK so flower power didn't work, so what? We start again."[23] Gazing at Rubin at the Nixon inauguration protest in January 1969, Hunter S. Thompson had already reflected that "we have come a long way from Berkeley and the Free Speech Movement. There is a new meanness on both sides . . . and no more humor."[24]

A key cause of this grim uneasiness was the increasing awareness of coordinated and intensified government surveillance of movement leaders, radical groups, and sympathetic celebrities. The 1968 Omnibus Crime Control and Safe Streets Act had made it possible for electronic surveillance, formerly a strategy of last resort reserved for national security measures, to become an ordinary means of law enforcement.[25] After the massive nationwide uprising protesting 1970's Cambodia invasion and subsequent Kent State killings, Nixon had quickly moved to "centralize domestic intelligence and counterinsurgency programs in his own hands." "John Mitchell's Justice Department," wrote the sociologist and activist Todd Gitlin, "with eager local assistance, pumped out special anti-radical grand juries, subpoenas, surveillances, wiretaps, campus spies, riot training, high-tech weapons. The army ran its own parallel operations, infiltrating a vast range of meetings and demonstrations, setting up dummy journalist teams, cooperating with right-wing vigilantes."[26]

The scope of these suppressive programs was decisively revealed in March 1971, when antiwar activists broke into the Media, Pennsylvania field office of the FBI, stealing documents that made public the existence of the COINTELPRO program. In its own words, COINTELPRO aimed to "expose, disrupt, and otherwise neutralize the activists of the 'New Left' by counterintelligence methods" and "enhance the paranoia endemic in [activist] circles."[27] Although this was achieved through all the strategies Gitlin names, the centrality of media technology—magnetic tape in particular—was of particular interest to the Firesign Theatre as they began recording *I Think We're All Bozos on This Bus* a few weeks later (they would also have appreciated that the break-in happened in Media). The protagonist's name, "Clem," seems to have come from the protagonist of William S. Burroughs's tape-themed *The Ticket That Exploded* (1962) and the cutup piece "Are You Tracking Me" (1965), and Firesign Theatre would have recognized that tape was (even more directly than it had been for Burroughs) their medium, too. This recognition, along with the COINTELPRO revelations, may have been one of the reasons they chose to abandon the trickery of tape edits that had made the storytelling of *How Can You Be* and *Dwarf* so distinctive.

Bozos abandoned the tape edit as a formal strategy—it was "our first razorblade-less album," said Ossman—and instead gave the technique a thematic meaning that it had not had on the earlier albums, making tape recording a phenomenon within the world of the story.[28] At the beginning of the album, as the ice cream bells fade out, the voice of a carnival barker announces the arrival of the bus sent to recruit visitors to the Future Fair:

PA VOICE: Live in the Future! . . . It's right around the corner. Yes, in fact, it's come right here to beautiful [tape insert] THIS AREA!

CLEM: What's that?

PA VOICE: . . . Join the expectant crowd gathering now, as we stop here on [tape insert] DUTCH ELM STREET.

This is the first indication that the Fair will be run entirely by nonhuman actors, speaking from what Weizenbaum's ELIZA program called a "script" keying in the appropriate (or inappropriate) information specific to a given location or situation (beautiful THIS AREA). The same technique is used to personalize users'/visitors' interactions with the president:

CHESTER: Maybe my friend would like to ask something. He's been with us since before the Beginning, sir. He must be bursting with a short question about his place in the Future. . . .

MR BROWN: Well, ask him his name.

CHESTER: Oh, I'm sorry. Could you state your name? [*Long pause*] Please state your name.

CLEM [*after a pause*]: Oh, do you want me to . . .

CHESTER: Please state your first name.

CLEM: Uh-Clem . . .

CHESTER: Thank you [tape insert, Clem's voice] [UHCLEM]. Now we can introduce you, and some of your selected neighbors, to our President.

And the recursive effect of this data collection, as well as its intended use for surveillance and control, is shown after Clem's first attempt to hack into the Fair, when we hear PA announcements on the midway attempting to locate and detain him: "Will our Guest Mr. AHCLEM please drop by the Hospitality Tower immediately?" The data may have been doubly corrupted (Clem → UHCLEM → AHCLEM) but the holograms are able to find him anyway.

It is worth stressing that the tape-mediated voice interactions of the Future Fair are the Firesign Theatre's improvised invention on Weizenbaum's ELIZA program, which would not have been accessed through voice commands but by typing questions on a keyboard and receiving printed responses spooled off a teletype machine. After interacting with ELIZA at the LA work fair, Proctor absconded with a sheaf of printouts from which the group drew their own script for *Bozos*.[29] Firesign fans later determined that the computer he would have found running ELIZA would have been a DEC PDP-10 (fig. 18), a mainframe computer used in many universities in the 1970s including MIT's AI Lab.[30] By joining ELIZA with the contemporary awareness of tape surveillance, they dramatized something close to the VUI (Voice User Interfaces) of the present day (as had HAL 9000 in Stanley Kubrick's 1968 film *2001: A Space Odyssey*). And Firesign also appreciated that computers such as the DEC PDP-10 used magnetic tape for storage, thus creating a material link from the world of early AI to the technology of surveillance and to their own recording practices at Columbia Square. When Clem asks the computer's "Dr. Memory" if it remembers the future, and the computer answers yes, Clem's final command is *"Forget it, tapehead!"*

Whenever the robots address a human character on *Bozos*, the message is personalized with a prerecording of the human's spoken name (Barney, Jim, Uh-Clem) and the album dramatizes the fact that it is a tape recording through the addition of production effects associated with tape: the warbling modulation of tape flutter, the speed variation produced by

Figure 18. DEC PDP-10 computer, manufactured by the Digital Equipment Corporation from 1966 to 1983. Fans later deduced that the Firesign Theatre interacted with the DEC PDP-10 running the Lisp version of ELIZA in 1970. DEC promotional brochure (1969). Courtesy of the Computer History Museum.

"reel-rocking" (the predecessor of digital "scrubbing"), and the EQ effect of a high-pass filter. These acoustic effects were made possible by new conditions in the rapidly evolving studio system. Columbia Square's Studio B now featured a sixteen-track recorder (as opposed to the eight-track machine used on *Dwarf* and *How Can You Be*, or *Waiting for the Electrician*'s four-track machine). This mitigated the need to edit or splice as a matter of practical exigency, reduced the need to bounce down into submixes, and afforded much greater precision for postproduction effects, doubled voices, and stereo mixing than before.

"There were enough tracks so you could keep everything and of course you didn't have to process anything," Ossman said. "You could keep everything flat and wonderful . . . until the mix down."[31] As a result, the overall sound of the album is quieter and deeper than the previous records; paired with the Dolby noise reduction technology introduced on *Dwarf*, this increased sonic control allowed for the ambient elements to be pulled further back in the mix, while remaining audible to the attentive (stoned) listener (Columbia released a quadraphonic mix of the album in March 1972).[32] As the track sheets show, the subdued opening of *Bozos* employs all sixteen channels, with separate tracks for footsteps, wind, hum of an

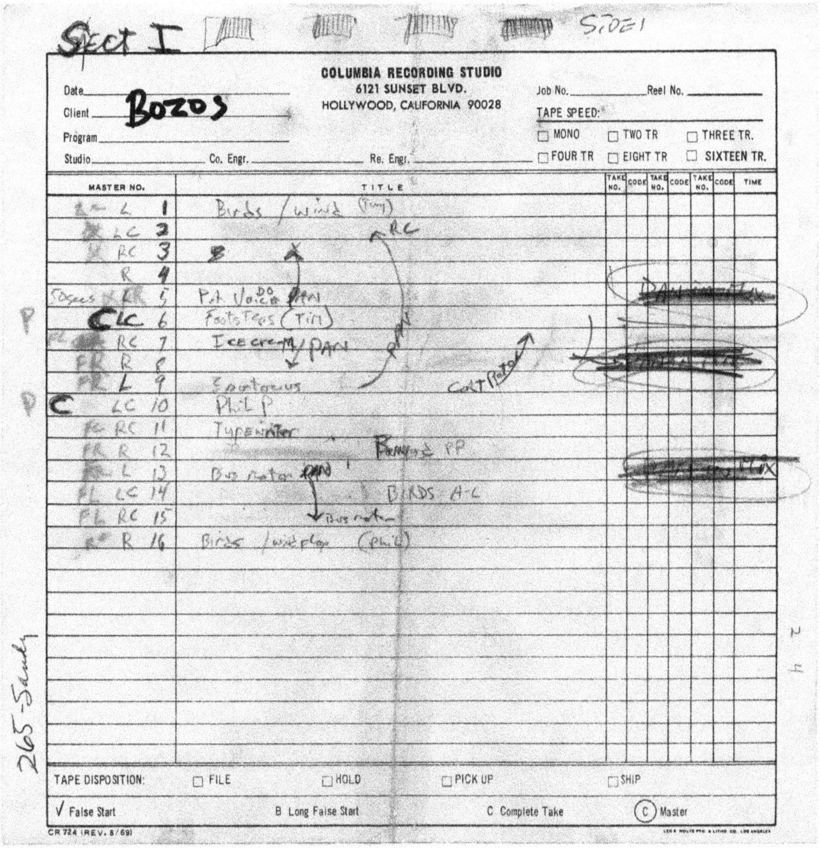

Sect I SIDE 1

COLUMBIA RECORDING STUDIO
6121 SUNSET BLVD.
HOLLYWOOD, CALIFORNIA 90028

Date
Client Bozos
Program
Studio Co. Engr. Re. Engr.
Job No. Reel No.
TAPE SPEED:
MONO TWO TR THREE TR.
FOUR TR EIGHT TR SIXTEEN TR.

MASTER NO.	TITLE	TAKE NO.	CODE	TAKE NO.	CODE	TAKE NO.	CODE	TIME
L 1	Birds / wind							
LC 2								
RC 3								
R 4								
L 5								
LC 6	Footsteps (Tin)							
RC 7	Ice cream / PAN							
R 8								
L 9								
LC 10	Phil P							
RC 11	Typewriter							
R 12								
L 13								
LC 14	BIRDS A-C							
RC 15								
R 16	Birds							

265-Sandy

TAPE DISPOSITION: FILE HOLD PICK UP SHIP
√ False Start B Long False Start C Complete Take (C) Master
CR 724 (REV. 5/69)

Figure 19. Studio sheets for the sixteen-track "quiet" opening of *I Think We're All Bozos on This Bus*, showing directions for bouncing tracks, stereo panning, and discrete tracks for footsteps, voices, wind, typewriter hum, two for the bus, two for the ice cream truck, and three for birds. Firesign Theatre Collection, National Audio-Visual Conservation Center, Library of Congress.

electric typewriter, two tracks for the approaching bus, two for voices, two for the receding ice cream truck, and three for birds (fig. 19).

This quieter, more controlled, sound world was of a piece with the warm, damp sound that characterized a wide range of rock and R&B records of the same year—*Tapestry*, *Hunky Dory*, *Meddle*, *What's Going On*, *Blue*, *Tago Mago*, *L.A. Woman*, *Electric Warrior*, *Master of Reality*, *Maggot Brain*, *There's a Riot Goin' On*—something that was likely an effect, also, of the solid-state transistor mixing boards that were replacing the tube-driven boards of previous decades.[33] But Firesign employed this warm effect to a very specific end. Although some of the other albums have rightly been read as paranoid classics—as in Marcus's famous account of *There's a Riot Goin' On*[34]—*Bozos'* conformity with the more sedate aesthetic of the year's rock recordings can itself be heard as thematic, as it pointedly cordoned off the topical references to political violence that could be heard somewhat more directly in the more hectic moments of *How Can You Be* and *Dwarf.*

That this was a deliberate choice is evident from the one exceptional eruption of violence into the story: when Clem is hailed on the Fair midway to enter an attraction (though possibly also a reality) called Mark Time's Outlaw Ghost Ship: "I need some brave boys and girls who aren't afraid to live outside the Law of Gravity. Families who like to sleep in tubes and push buttons. Adventurers like you." The likely reality of the journey is betrayed by Mark Time's android companion, who shouts, "Mark! Mark! Mark! They're performing experiments on animals in space! Mark! Mark! Mark! They're performing horrible . . ." before Mark kicks the robot back into submissive compliance. This can be heard as dramatizing a form of suppression in the way that *Dwarf* had allegorized the political firing of the Black DJ Roland Young from KSAN. By contrast, the Fair's mode of mandatory complacency is emblematized by the animatronic vegetables who accost Clem at the beginning and stage a mild squabble before the Whisperin' Squash admonishes, "Now, now, boys—fightin's out of style."

MAN CONFORMS

There would have been other ways to represent the disciplinary control of dissent in public life and the rise of the surveillance state. But choosing the administered, technophilic environment of a theme park as the album's allegorical figure allowed the Firesign Theatre to emphasize the way social control could be naturalized as a form of what is now called "user engagement." In his influential 1969 synthesis, historian Theodore Roszak had proposed that the militant New Left and the less obviously political hippies could together be seen as reactions to what he called "the technocratic society," a world of "social engineering in which entrepreneurial talent broadens its province [beyond the factory] to orchestrate the total human context which surrounds the industrial complex. Politics, education, leisure, entertainment, culture as a whole, the unconscious drives, and even . . . protest against the technocracy itself: all these become the subjects of purely technical scrutiny and of purely technical manipulation."[35] In the world of *Bozos*, the Future Fair is both an allegory of and a mechanism for producing this "total human context." Firesign must have been at least ambiently aware that the carnivalesque Renaissance Pleasure Faire, as a utopian critique of the technocracy, had been the Future Fair's opposite.

Disneyland would have been an obvious source for this critical allegory, not least because of its proximity to Los Angeles and its deep ties to President Richard Nixon. Nixon had grown up in Yorba Linda and had been a welcome guest at Disneyland after it opened in 1955. White House Chief of Staff H. R. Haldeman had been an executive at an advertising agency that represented Disney, and Press Secretary Ronald Ziegler had spent his college years operating a boat on Disneyland's Jungle Cruise ride.[36] Yet although elements of Disneyland are present in mediated form throughout the album—particularly the audio-animatronic president—Firesign's technocratic critique was much better served by the source they drew on most strongly—the 1933 Century of Progress World's Fair.

Unlike the medieval carnivals emulated by the Pleasure Faire, the phenomenon of the world's fair emerged in the age of industrialization as a "device for the enhancement of trade, for the promotion of a new technology, of the education of the ignorant middle classes and for the elaboration of a political stance."[37] Though always an instrument of official culture, there were reasons that the Chicago Century of Progress exposition was especially useful for *Bozos*. Besides eliminating the notorious racist practice of human display that had marred fairs for decades (which was actual progress), the 1933 fair was the first to abandon classical architecture and valorize progress with the utopian language of modernism, and it also featured an unprecedented collaboration with private industry (in this way anticipating LACMA's Art and Technology program).[38]

In her history, Cheryl R. Ganz describes A Century of Progress as "a civil-military enterprise [that] reinvented the concept of international expositions" through the new concept of thematic halls and corporate pavilions, along with futuristic buildings for Home and Industrial Arts,

Figure 20. Postcards from the 1933 Century of Progress World's Fair, showing the midway's "Bozo" ride, the "intra-mural" bus in front of the Hall of Science, and the carillon tower of the Hall of Science. Collection of the author.

Radio and Communications, Travel and Transport, Homes of Tomorrow, and the US Government interspersed among those for Chrysler Motors, General Motors, Sears Roebuck, Time-Fortune, and Sinclair Oil. Together, these "replaced orthodox [humanist] views with their belief that progress rides on the swell of technological innovation."[39] The fair's thematic slogan was "Science Finds—Industry Applies—Man Conforms." Shoshana Zuboff would cite this motto eight decades later in the introduction to her landmark book *The Age of Surveillance Capitalism.*[40] Like the Firesign Theatre, Zuboff recognized that the media archaeology of artificial intelligence led backward to the techno-utopianism of the world's fair.

Firesign borrowed both the term *Bozo* ("a midway ride") and *Bus* ("especially built by General Motors") from the *Official Guide Book.* But the album's primary object of extemporization was the fair's central pavil-

Figure 21. Louise Lentz Woodruff's *Science Advancing Mankind* statue at the Hall of Science, Century of Progress World's Fair, Chicago 1933.

ion, the Hall of Science, which extended over eight acres beneath a 176-foot-high carillon tower (fig. 20). Portraying scientific research as the servant and engine of corporate industry, the Hall of Science dramatized historical "torch-bearers of science," the physical sciences, new technologies like neon light, developments in medicine (including "the world's most beautiful drug store"), and "the romance of gas and oil."[41] At the heart of the pavilion was Louise Lentz Woodruff's striking and peculiar sculpture *Science Advancing Mankind*, which comprised a giant cubist robot urging forward two passive figures, arms blindly extended, representing humanity (fig. 21). In Ganz's synopsis, "science and technology, independent of human agency, drive progress" (3); a 1952 remembrance dubbed it "Technology's Triumph Over Man."[42]

In the first minutes of *Bozos*, Clem boards the bus that takes him to the Future Fair, where he is one of the few passengers who is not a "Bozo" (his neighbor Barney has inflatable shoes and offers to let Clem squeeze his nose). Disembarking near the fair's Mindless Fellowship Pavilion, Clem is instructed to "clone under the Big Blue 'B,' up against the Wall of Science." This phrase proves to be a multivalent riff on the Hall of Science's corporate triumphalism. The substitution of Wall for Hall (a boundary for a pathway) is obvious, but for Clem to be exhorted "up against the Wall of Science"—a phrase that the exhibit itself repeats—seems a more specific reference to the human figures in Woodruff's statue, whose outstretched arms would have resembled, in 1971, belligerents being pushed "up against the wall" by the military or police. And the pun cuts the other way as well. Just as *How Can You Be* had riffed on the SDS slogan "bring the war home," "up against the wall of science" namechecked the "Up Against the Wall Motherfuckers" (or simply "Motherfuckers"), the Lower East Side "street gang with analysis," which formed in the late 1960s, took its name from Amiri Baraka's poem "Black People," and participated in the 1968 student takeover at Columbia University (where they dressed as a fife-and-drum corps).[43] In either reading of "up against the wall of science," the technocracy is the terrain of combat.

Firesign's Wall of Science combined elements from the Hall of Science and the Sinclair Oil pavilion "The World—A Million Years Ago," which showcased mechanical dinosaurs that "roamed o'er Mother Earth . . . while Nature was mellowing the crudes that today are used to refine the motor oil sold under this company's well-known brand names" and featured a moving sidewalk that transported visitors past the exhibits.[44] Listeners to *I Think We're All Bozos on This Bus* are figured as a similarly mobile audience, sharing the same imaginary space in the story as Clem, who at this point recedes in the narrative. They are then immersed in a twelve-minute sequence of aural dioramas that together portray a comically grand and distended temporal drama that stretches from the creation of the earth; the emergence of plant, animal, and human life; human development from primitive society, through early scientific discovery, into modernity; and at last culminates in the creation of a "new model government" in the present.

It is one of the most elaborate soundscapes in the Firesign Theatre's entire catalog, each of the sixteen discrete installations being defined by its own distinct aural design. In addition to the voice work of the actors, each minienvironment is given its own array of ambient effects (Proctor produced the effect of thunder by beating on the giant spring in Columbia Square's reverb chamber); voices are treated with postproduction effects like tape echo, doubling, and phasing; some are recorded in deliberately poor audile environments, as in the short sequences meant to depict college lecture halls (complete with coughing, chairs shifting, and speaking off-mic to indicate the distance of the TA at the chalkboard).[45]

The story, before the advent of humanity, is told by a succession of mock-authoritative voices, from the Voice of God ("Before the beginning, there was this turtle . . .") through a sequence of self-certain scientists:

> We know for certain, for instance, that for some reason, for some time in the beginning, there were hot lumps. Cold and lonely, they whirled together noiselessly through the black holes of space. These insignificant lumps came together to form the first union—our Sun, the heating system. And about this glowing gasbag rotated the Earth, a cat's-eye among aggies, blinking in astonishment across the Face of Time. . . .
>
> Animals without backbones hid from each other or fell down. Clamasaurs and oysterettes appeared as appetizers. Then came the sponges, which sucked up about ten percent of all life. Hundreds of years later, in the Late Devouring Period, fish became obnoxious. Trailerbikes, chiggerbites and mosquitoes collided aimlessly in the dense gas. Finally, tiny edible plants sprang up in rows, giving birth to generations of insecticides and other small, dying creatures.

Woven through these short chapters is a joke about the patriarchal monopolization of knowledge. The originary myth posits a giant walking catfish with two balls, "one of which is the Sun, and the other they called the Moon." These are demystified by the expert ("some uncomplicated peoples still believe this myth"), who recognizes the balls as the "hot lumps" that would form the Sun.

At length, the clamasaurs and oysterettes are displaced by the "modifying spark of humanity," and the evolutionary narrative culminates in a diorama that depicts the jejune nineteenth-century scientists of the New England Amateur Electrical League, who "put the balls on the other side" of their machine and discover Teslacle's Deviant to Fudd's Law. "Those balls will mean your fortune!" Now, "with the invention of the Motor Operated Pushover, Man and Science gave birth to life here, today, in the Future! Man, woman, child! All is up against the Wall of Science!" That mastery, in turn, becomes the precondition for the final animatronic diorama, in which the new model government is activated when it hears the spoken word "Power" (uncannily anticipating contemporary VUIs like Siri and Alexa), which in turn leads to the Wall of Science's central attraction—not a statue of Science Advancing Mankind but the robotic president himself.

YOU HAVE VIOLATED ROBOT'S RULES OF ORDER AND WILL BE ASKED TO LEAVE THE FUTURE IMMEDIATELY

The sequence with the president, which involves Clem's return as an active character, draws on Disneyland as well as the World's Fairs, adding to these references inferences drawn from the stolen ELIZA printouts. Much of the dramaturgy of the Future Fair anticipates Eco's 1975 essay "Travels in Hyperreality," which begins with a visit to New York's School of Holography and then spends several pages meditating on the "fantasy world more real than reality," the "total theater" of Disneyland. In particular, Eco is struck by Disneyland's speaking and moving Audio-Animatronic robots, "master-pieces of electronics . . . authentic computers in human form, . . . dressed in 'flesh' and 'skin' made by craftsmen, whose command of realism is incredible. Each robot obeys a program, . . . repeating ad infinitum all day long his established part."[46] The Future Fair of *Bozos* is administered by characters that identify themselves as holy-grams, and the nearest referent for the speaking characters of the Wall of Science would be Disney's robots, each of which repeated its lines via loops of prerecorded audiotape.

Disney's tech workers—their official title was, and remains, "Imagineer"—had in fact already made an Audio-Animatronic president (Abraham Lincoln) for the 1964–65 World's Fair in New York. It was then installed at Disneyland, where visitors could hear it speak excerpts from several of Lincoln's speeches.[47] Improving on the robot Lincoln, the Wall of Science's Mr. President is able to interact with fairgoers—to the point of answering a "state[d] short question"—and in this way represented something more than a straight copy of the robots that so astonished Umberto Eco. Rather, it imagined at the level of sound the computer interactions that the ELIZA program had pioneered at the level of input text. Firesign imagined what would indeed prove to be the case: ELIZA would be the forerunner of contemporary VUIs.

After finishing at Wayne State, Weizenbaum moved to what would later be called Silicon Valley, working as a programmer for General Electric, where he was contracted to build a computer for the Navy that launched missiles and a check-processing computer for Bank of America.[48] But in 1963, he was invited to join the faculty at MIT's Project MAC, working to develop the emergent "time-sharing" technology that allowed a more conversational form of interaction with computers (MAC referred to Multiple Access Computing, but to the MIT's AI Lab, it came to mean Machine Aided Cognition). Weizenbaum developed the program ELIZA between 1964 and 1966; it was an early natural-language-processing program that used a script (one of Weizenbaum's many theatrical metaphors) to simulate conversation through a "multistep transformation of the most recent audience statement."[49] The script "decomposed" an input statement from the human user, performed simple substitutions (I for YOU), and searched for key words to produce a plausible response:

> You are not very aggressive but I think you don't want me to notice that.
>
> WHAT MAKES YOU THINK I AM NOT VERY AGGRESSIVE
>
> You don't argue with me.
>
> WHY DO YOU THINK I DON'T ARGUE WITH YOU[50]

According to the script Weizenbaum included as an appendix to his first 1966 essay on ELIZA, the code producing the first response would process the phrase "You are not very aggressive" according to the code (0 I ARE YOU) (WHAT MAKES YOU THINK I AM 4). Weizenbaum named the program ELIZA after the character in Shaw's play *Pygmalion*, with the idea that it might be "incrementally improved by its users."[51] He wrote that from one perspective, the script was a program and ELIZA itself an interpreter, but from another perspective, "ELIZA appears as an actor who must depend on a script for his [*sic*] lines."[52]

Any number of scripts could be written using the ELIZA program, but by far the most well-known script was DOCTOR, in which ELIZA played the role of a nondirective Rogerian psychoanalyst, who consistently put the burden of interpretation back on the analysand:

> Men are all alike.
>
> IN WHAT WAY
>
> They're always bugging us about something or other.
>
> CAN YOU THINK OF A SPECIFIC EXAMPLE
>
> Well, my boyfriend made me come here.
>
> YOUR BOYFRIEND MADE YOU COME HERE . . .[53]

In the words of theater scholar Christopher Grobe, ELIZA-DOCTOR was a "jury-rigged mixture of keyword recognition and stalling tactics."[54] Indeed, Weizenbaum stressed that "one of the principle [*sic*] aims of the DOCTOR program is to keep the conversation going—even at the price of having to conceal any misunderstandings on its own part."[55]

This was a feature that the Firesign Theatre reversed when they orchestrated an ELIZA-like interaction for the Future Fair. Substituting the president for DOCTOR's psychotherapist, the Future Fair's exhibit presents the fantasy of responsive government; you can ask the president a question, but it makes sure that it is only one question. Foreclosing the possibility of conversation also helps to conceal the computer president's lack of understanding (more accurately, the political system's lack of caring), and on first glance, it would appear that Firesign's point is to allegorize the "scripted" quality of all public appearances of political figures. The real president's TV consultant, Harry Treleaven, had proudly stated in 1968 that "Nixon has not only developed the use of the platitude, he's raised it to an art form."[56] Firesign dramatized this banality by voicing Phil Austin's Mr. President through the rotating Leslie speaker of the Columbia Square Hammond organ (the one effect that was applied live in tracking, so that Austin could "play" it as a kind of instrument as he spoke).[57] John Lennon had sung through a Leslie speaker on the Beatles' "Tomorrow Never Knows," hoping to sound like "the Dalai Lama singing from the highest mountain top."[58] The shuddering vocality of Austin's Nixonesque president has the opposite effect on its authority.

The ELIZA phenomenon provided the Firesign Theatre with a model for presenting political speech as a form of automated, "on message" doublespeak, and the president's response to each of the three characters who engage him can be seen according to these terms. To any listener's ears, each of the president's platitudes is easily deconstructible, which was part of the album's dark humor. When the first fairgoer, a Black American named Jim, angrily asks the president, "Where can I get a job?" the president responds: "M-many busy executives ask me, 'What about the Job Displacement Market Program in the City of the Future?' Well, count on us to be there, [tape insert] JIM. Because, if we're lucky tomorrow, we won't have to deal with questions like yours ever again." The self-deconstructing joke here is, of course, that the jobs program is really a "Jobs Displacement" program and the president's true wish is that he "won't have to deal with questions" like the ones he's asked "ever again." (Before he's even finished, Jim mutters, "He's jiving me again.")

What is notable, though, is that the president's responses are not simply key-word based, as they would be in an ELIZA-based script (where can I get a job → Jobs Displacement Market Program → count on us to be there). They also seem to be produced by the racially marked voices of the president exhibit's discrete patrons, who are (in order) the African American character Jim, the white Bozo Barney (first seen on the bus next to Clem), Clem, and an unnamed Latino character. The subtly variegated responses each of these characters receive suggest that the Firesign Theatre were also intuiting something closer to the problematics now evident in "machine listening," which, unlike the text-based ELIZA, through voice recognition software, analyzes the voice for "a particular emotion, language, race, or gender, or a certain degree of truthfulness" and constructs profiles based on those presumptions.[59]

The duplicitous and patronizing response Jim receives may be key-word based, but it differs substantially in tone from the one Clem receives after his first attempt to hack into the computer:

CLEM: This is Worker speaking. Hello.

PRESIDENT: Hello. How are ya? State maintenance question.

Clem does not even ask a civics question—he tries to destroy the computer by asking it unanswerable nonsense—but when the president program recovers (or reboots), it addresses Clem in a tone of sympathetic identification (maybe as another white professional) that it denies Jim: "And I'm as tired of it as you are. And I hope that our children will come to love us again in some better world than this." That these subtly differentiated attitudes are produced by racialized machine listening is also emphasized by the president's preprogrammed salutation of the fairgoer, each of which seems to have something to do with time. The greeting Jim receives is redolent of social tension, "Always nice to see you [tape insert] JIM. You know, the mainspring of this country, wound up as tight as it is, is guaranteed for the life of the watch." But the Bozo Barney is given a platitude that wanly celebrates the idea of human progress: "Always glad to talk to you, [tape insert] BARNEY. You know, when you clock the human race with the stopwatch of History, it's a new record every time." (The latter would be the first Firesign phrase to be sampled in a hip-hop track, by Mark the 45 King on Chill Rob G's "The Power" in 1989.) If understanding this embedded critique does not excuse Bergman's blackvoice performance as Jim, it does at least explain it.

Firesign did not only parody the ELIZA-DOCTOR script, sometimes quoting directly ("I am not sure I understand you fully"), but they also included in their writing technical noise words that came not from the Lisp language that made up the ELIZA program but from the operating system of the DEC PDP-10 computer Proctor had interacted with at the work fair (fig. 22). These were the words they wove into the script to dramatize Clem's hacking through the president into the back end of the computer, giving Firesign's script an authenticity that undoubtedly led to the album's later cult status in Silicon Valley.

CLEM: Read me Dr. Memory.
PRES: AMRAD . . . It's not sure I understand you fully. Could you state that as a question, please?
CLEM: Read me Dr. Memory?
PRES: SYSTAT. DIRECT READOUT. UPTIME 9:01 . . . I have been awake for 9 hours, 1 minute, 44 seconds. AMYL FAX SHUFFLE TIME. Less than 1 percent of freight drain, LOG FIVE. 5 jobs, 2 detached. MINIMUM ENTRY. GATE ONE . . .

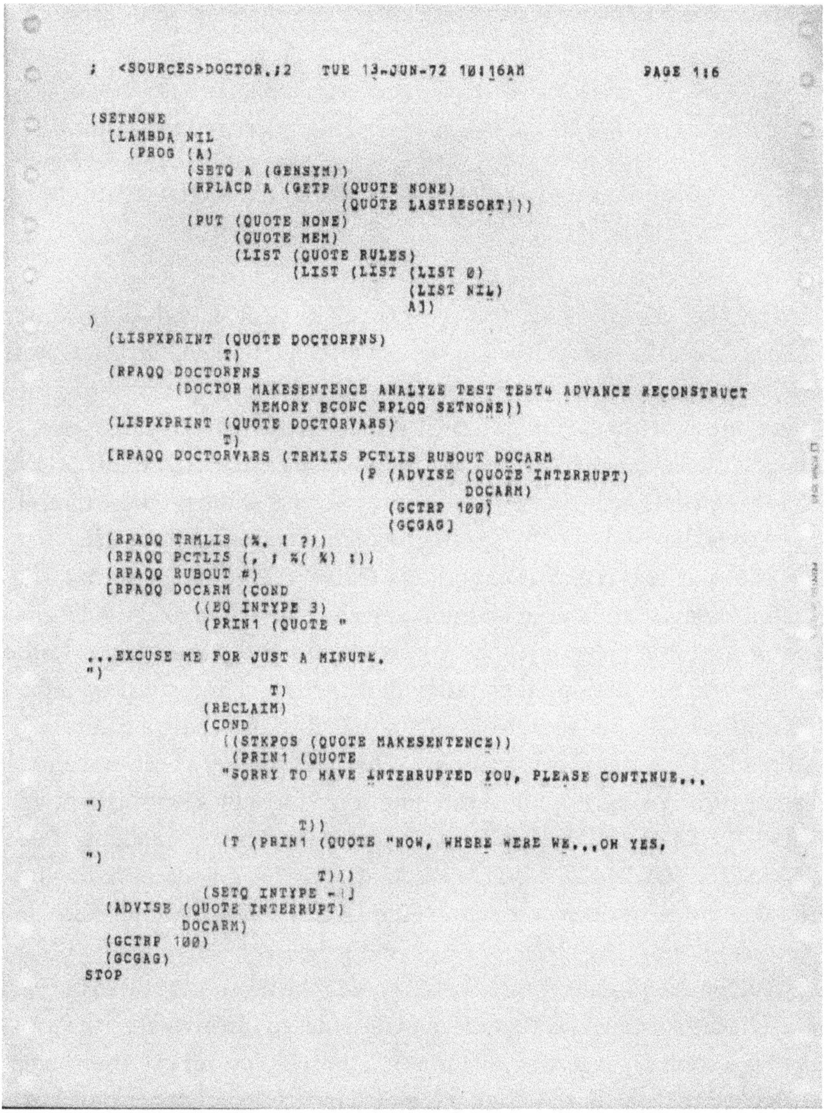

```
(SETNONE
  [LAMBDA NIL
    (PROG (A)
      (SETQ A (GENSYM))
      (RPLACD A (GETP (QUOTE NONE)
                      (QUOTE LASTRESORT)))
      (PUT (QUOTE NONE)
           (QUOTE MEM)
           (LIST (QUOTE RULES)
                 (LIST (LIST (LIST 0)
                             (LIST NIL)
                             A])
)
(LISPXPRINT (QUOTE DOCTORFNS)
            T)
(RPAQQ DOCTORFNS
       (DOCTOR MAKESENTENCE ANALYZE TEST TEST4 ADVANCE RECONSTRUCT
               MEMORY ECONC RPLQQ SETNONE))
(LISPXPRINT (QUOTE DOCTORVARS)
            T)
[RPAQQ DOCTORVARS (TRMLIS PCTLIS RUBOUT DOCARM
                   (P (ADVISE (QUOTE INTERRUPT)
                              DOCARM)
                      (GCTRP 100)
                      (GCGAG)]
(RPAQQ TRMLIS (%. ! ?))
(RPAQQ PCTLIS (, ; %( %) :))
(RPAQQ RUBOUT #)
[RPAQQ DOCARM (COND
               ((EQ INTYPE 3)
                (PRIN1 (QUOTE "

...EXCUSE ME FOR JUST A MINUTE.
")
                       T)
                (RECLAIM)
                (COND
                 ((STKPOS (QUOTE MAKESENTENCE))
                  (PRIN1 (QUOTE
                  "SORRY TO HAVE INTERRUPTED YOU, PLEASE CONTINUE...

")
                         T))
                 (T (PRIN1 (QUOTE "NOW, WHERE WERE WE...OH YES,
")
                           T)))
               (SETQ INTYPE -1]
(ADVISE (QUOTE INTERRUPT)
        DOCARM)
(GCTRP 100)
(GCGAG)
STOP
```

Figure 22. Eliza/Doctor in Lisp by Bernie Cosell, June 13, 1972. Courtesy of Jeff Shrager, ELIZAgen.org.

CLEM: Alright, you're doing fine, but this is a Flip-Flop, Springhead.

PRES: FLIP-FLOP . . .

CLEM: I'm gonna ask you a question that you won't be able to answer.

PRES: I am not sure I understand you fully. LOG OUT. RUN OFF. MEMORY. A.—"The system is less energetic if domains of opposite directions alternate." NPN READ MACNAM. PNP READ MACNAM

What the script does not convey—but can be easily heard—is that once Clem accesses the system, the president's speech is broken into three vocal performances, Ossman and Bergman joining Austin to perform together something closer to the mainframe itself, which they named "Dr. Memory." Along with the pastiche of spoken code ("SIS, PUP, DAD, BUB, MOM, PIP, DAD, DUD"), the three voices convey a sense of the multiple discrete operations that underlie the president's official, automated speech.

The sequence is remarkable both as a feat of writing and performance—Austin, Ossman, and Bergman hum and whirr as well as read the script—and of sound engineering. The voices warble both through the Leslie speaker and with the speed variation of reel-rocking and are close-miked to foreground deeper vocal frequencies (possibly also using a compressor or limiter). Clem's goal is to gain access to the computer's back end and to disable it by asking it a nonsense question. It briefly seems as if it will work—"LAUGH. RUNAWAY. The doctor makes no readout. Read UNHAPPY MACNAM. UNHAPPY MACNAM"—and succeeds in shutting down the president attraction (the Latino character, under no illusions about what the attraction is, is nevertheless disappointed: "I been waiting for ten minutes from here to listen to the president talk to me"). It is significant, meanwhile, that the fair continues to run even after the president is disabled, suggesting that the true source of power lies elsewhere.

Departing to the Funway, Clem is hailed by the PA announcements that attempt to draw him to the disciplinary "Hospitality Shelter." Accosted by the vegetable hologram Artie Choke, he hacks in again ("This is Worker speaking. Hello." "Hi. Open for maintenance alignment.") and this time accesses the mainframe of the computer in a way that resembles Kevin Roose's alarming experiments with the Microsoft AI chatbot in February 2023, a salient act of "AI red teaming" that began with Roose compelling the chatbot to reveal its secret name, "Sydney."[60] Similarly, the Future Fair's president is subtended by the occluded name Dr. Memory. Rather than mimicking the ELIZA printouts here, Dr. Memory's verbal "noise" operates along the lines of ELIZA's primitive grammar (HE SHE IT), but according to a more sophisticated, sound-associative or punning principle that responds to the sound of Clem's words—which so far are unable to articulate a fully formed question—instead of the thematic "key-word" logic that had governed the president's responses. This associative function becomes especially clear when Clem is able to ask the question that will make the entire system vulnerable and bring it to the point of collapse: "Why does the porridge bird lay his egg in the air?"

Because it assumes as given something that can only be true fictively—having the structure of a logical question but which is in the strictest sense "irrational"—Clem's Jabberwockish question defies the possibility of Dr. Memory's providing a plausible response, eliciting instead a string of punning restatements of the question:

CLEM: All right, Doc—this is it! Gird your grid for the big one! Why does the porridge bird lay his egg in the air?

DR. M: NOOOOOOOOOOOOOO...

MAC 2:   White dust 'n' perished birds leaves its hex in the air?

DR. M: NOOOOOOOOOOOOOO...

MAC 3:   Wise doves 'n' parish bards lazy leg in the Eire?

DR. M: NOOOOOOOOOOOOOO...

MAC 2:   Wise ass the poor rich Bar [*Honk*] ...

MAC 3:   DELAY...

MAC 2:   ... laser's edge in the Fair?

DR. M: NOOOOOOOOOOOOOO...

MAC 3:   The Doctor is unhappy...

When Clem presses the computer about whether it can answer his question, the computer answers simultaneously no and yes; a third voice interrupts, "Excuse me/don't excuse me, Uh Clem, you're making/not making the Doctor unhappy/happy." Clem then poses the questions that underpin the computer's orchestration of the entire social reality of the fair:

CLEM: Do you remember the Past, Doctor?

DR. M: YESSSSSSSSSSSS . . .

CLEM: Do you remember the Future?

DR. M: YESSSSSSSSSSSS . . .

Clem then issues the command, "Forget it!," causing the machine to explode, returning the listener to the desolate world of possibility where the record began, outside the fair.

In a work distinct among their albums (to date) for its employment of linear narrative, Firesign Theatre nevertheless claimed a heroic role for the irrational. If, as Marcus argued, this was an "overly rational" employment of the irrational, it was also a brief for the group's faith in its political agency and was associated with the creative power of the aesthetic. "I here declare the end of the War!" proclaimed Allen Ginsberg in his long poem "Wichita Vortex Sutra," which named the media and technology companies profiting from Vietnam and pretended to defeat them with his bardic verse.[61] In his later, heretical writing, Weizenbaum quoted the absurdist playwright Eugène Ionesco in a similar spirit: "Not everything is unsayable in words, only the living truth."[62]

ART AND TECHNOLOGY

This concatenation of issues was similar to those that led Joseph Weizenbaum to renounce his earlier work and, by the mid-1970s, become one of the first and most voluble opponents of the entire concept of artificial intelligence. It is striking to see the degree to which Weizenbaum's later heresy is prefigured in the dramaturgy of *Bozos*. Of the first three visitors who interact with the president, the Bozo Barney is the most excited to see the president, and he has by far the least satisfactory experience. The program hiccups before Barney is able to ask his question (maybe because he says the word "stop"), and the president merely thanks Barney and tells him he'll receive "a handsome simulfax copy of your own words in the mail soon. And my reply." Barney, however, is both unperturbed and embarrassingly grateful ("By golly, Mr. President, you're beneficent.") Though it had not been widely recognized at the time, Barney's response can be heard as a rehearsal of what became the most famous and oft-discussed secondary effect of Weizenbaum's invention: the so-called "ELIZA effect."

Even while he was diligently working on a program with goals clearly aligned with contemporary phenomena like Siri and Alexa—and the more recent, much heralded (and for now text-based) ChatGPT—Weizenbaum wanted to make clear that the computer responses to human input should not be confused with the authority of human understanding, and he tried to "rob ELIZA of the aura of magic to which its application to psychological subject matter has to some extent contributed."[63] What he did not anticipate was the degree to which human users would be attracted to cede this authority to machines. This confusion, encouraged by the engaging fun of the interface, has been the subject of the many subsequent studies of ELIZA that have been written since the 1990s. Few if any of those studies have taken up the existential concerns Weizenbaum began to raise as homegrown versions of the ELIZA program attained the status of what its inventor would describe as a "national plaything."[64]

Those concerns were all born of observations Weizenbaum made in the context of the convergence of "the civil rights movement, the war in Vietnam, and MIT's role in weapons development" (Project MAC had been established in 1963 with a grant from the government's Defense Advanced Research Project Agency, or DARPA).[65] In his 1976 book, *Computer Power and Human Reason*, Weizenbaum does not primarily worry about the individual users of ELIZA (as with the oft-told story of his secretary asking for privacy while she interacted with the program). Rather, he worries about the way it was uncritically received within the scientific community as a "general solution to the problem of computer understanding of natural language."[66] Weizenbaum recognized in these responses (from those who should know better) a general tendency to equate rationality with logic, a confusion he attributed to the increasing mechanization of humanity itself. He laid out his case as a dark revision of Marshall McLuhan's influential theory of media as "extensions of the senses":

> Western man's entire milieu is now pervaded by complex technological extensions of his every functional capacity. Being the enormously adaptive animal he is, man has been able to accept as authentically natural (that is, as given by nature) such technological bases for his relationship to himself, for his identity. Perhaps this helps to explain why he does not question the

> appropriateness of investing his most private feelings in a computer. But then, such an explanation would also suggest that the computing machine represents merely an extreme extrapolation of a much more general techno-logical usurpation of man's capacity to act as an autonomous agent in giving meaning to his world. It is therefore important to inquire into the wider senses in which man has come to yield his own autonomy to a world viewed as machine. (9)

He concludes his prefatory comments by asserting that "all empirical science is an elaborate structure built on piles that are anchored . . . on the shifting sand of fallible human judgment, conjecture, and intuition," hence the bleak trajectory implied in the book's subtitle, "from judgment to calculation" (14–15).

Himself a Jewish refugee from Nazi Germany, Weizenbaum writes of being "more than a little frightened" as he reflected on the times he conferred with MIT colleagues "on research proposals to be made to government agencies, especially the Department of [scare quotes] 'Defense'" (9). And he went further to declaim that "in the recent American war . . . computers operated by officers who had not the slightest idea what went on in their machines effectively chose which hamlets were to be bombed and what zones had a sufficient density of Viet Cong to be 'legitimately' declared free-fire zones, that is, large geographical areas in which pilots had the 'right' to kill every living thing. Of course, only 'machine-readable' data, that is, largely targeting information coming from other computers, could enter these machines" (238).[67] The degree to which this statement anticipates our contemporary situation, in which we have accepted not only these techniques but also the situation of war without end, is unfortunately very obvious. Weizenbaum was issuing these alarms from a tenured position in the university that had received by far the greatest funding from the Pentagon. As a colleague in information science put it to me candidly, "The call was coming from inside the house."

In the *Rolling Stone* profile, Peter Bergman had stated that "America put on a uniform in 1941—and hasn't taken it off since."[68] With *Bozos*, the Firesign Theatre waged symbolic battle (in their technologized art) against the technocracy generally but against the specific complicity of the ELIZA effect ("the world viewed as machine"). Such self-reflexivity was notoriously absent at LACMA's "Art and Technology" exhibition, which opened while

Firesign was recording *I Think We're All Bozos on This Bus* in Hollywood. That exhibition had paired cutting-edge artists with Southern California technology corporations to produce monumental works that would "not only complement the futuristic setting of Los Angeles but would also enhance the museum's financial and cultural standing."[69] Early results of the collaborations had appeared at the 1970 World's Fair in Osaka, sponsored by the United States Information Agency (USIA) (see chapter 2).

By the time of the 1971 LACMA exhibition, only sixteen of the seventy-nine partnerships had come to fruition. Those included R. B. Kitaj's *Lives of the Engineers* (a collaboration with Lockheed), Robert Rauschenberg's *Mud Muse* (Teledyne), Newton Harrison's *Plasma Tubes* (Jet Propulsion Laboratory), Andy Warhol's holographic *Rain Machine* (Cowles Corporation), and Claes Oldenburg's animatronic *Icebag* (originally conceived with Walt Disney Enterprises but ultimately fabricated by Sid and Marty Krofft!). As the critic Michael Fallon has since observed, the exclusively white and male roster of artists led to a counterinsurgency of identitarian art in Los Angeles, which largely rejected the new technologies as resources for art.[70] More immediately, "Art and Technology" was pilloried for what the critic Max Kozloff called the "rogue's gallery of the violence industries," its other partners including the Rand Corporation, the Garrett Corporation (which made jet engines for military aircraft), and Litton Industries (builders of submarines). "The American artists," wrote Kozloff, "did not hesitate to freeload at the trough of that techno-fascism that had inspired them."[71] Forty years later, in 2013, LACMA announced the launch of a new Art + Technology Lab, inspired by the 1971 exhibition, boasting the support of corporate sponsors Accenture, NVIDIA, DAQRI, SpaceX, and Google, and with an advisory board featuring an executive from the Jet Propulsion Laboratory.[72]

UNHAPPY MACNAM

The achieved "breaking" of the president, and subsequent destruction of the totalitarian world of the Future Fair, have been subsequently seen, by fans and by the group itself, as marking the apex and culmination of the group's "first four classic albums" period. Though I dissent from this

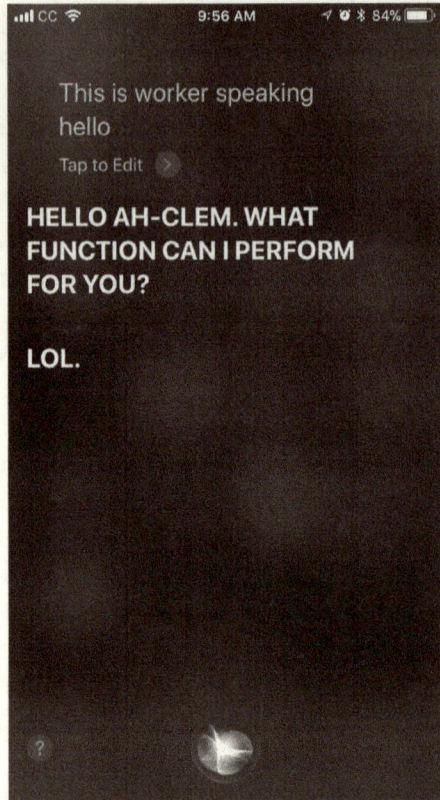

Figure 23. Screenshot of iPhone's
Siri responding to Clem's words from
I Think We're All Bozos on This Bus,
September 28, 2017.

opinion, there is no doubt that *I Think We're All Bozos on This Bus* marked
the end of one phase of the group's work and was fraught with ambiguity,
for reasons to be explored more fully in the next chapter.

Part of this ambiguity, however, can be explored by referring to the later
reception of the album. When Apple integrated its "intelligent personal
assistant" Siri into the iPhone in October 2011, it included a hidden message that referred directly to *Bozos,* and it remained an Easter egg through
the release of iOS 15 in 2021. When addressed with the words Clem uses
to access the Fair's infrastructure—"This is worker speaking"—Siri would
provide the knowing response, "HELLO AH-CLEM. WHAT FUNCTION CAN I
PERFORM FOR YOU? LOL" (fig. 23). Nor was this an isolated instance. When
asked Clem's disabling question, "Why does the porridge bird lay his egg

in the air?," the Siri user was referred to a separate AI site, Wolfram Alpha, which answered, "Nice try. While I don't know, you cannot get me to shut down that easily!"

These responses have finally been disabled, but they speak to a longer history of Firesign Theatre references in the culture of Silicon Valley, and at Apple in particular. As early as 1980, there was a word-processing program for Apple II called Dr. Memory. And in 1993 Eric Malone wrote to the *Four-Alarm FIRESIGNal* fanzine (from Mountain View, California) about a program called NetWare: "When installing the software, you have to identify what is known as the Macintosh Name Space so that other computers (DOS and UNIX) can recognize the computer and its files. The identification file is known as MAC.NAM. . . . When installing the file, we had a problem and the screen displayed this message: READ UNHAPPY MAC.NAM."[73] Philip Proctor is fond of telling the story of meeting Steve Jobs in 1998, who told him he had been inspired by *Bozos*.[74] Jobs is unlikely to have authored any of the aforementioned *Bozos* references, but their existence is evidence of a powerful strain of Silicon Valley's self-authentication, an economic libertarianism with countercultural roots that Richard Barbrook and Andy Cameron named, in their withering 1995 critique of *Wired* magazine's techno-utopianism, "the Californian Ideology." This line of self-fashioning was continuous with the heroic role for the figure of the hacker, a direct line from which can be drawn the dubious hypostatization of "disruption" as a cultural paradigm (celebrated annually by the plutocratic carnivalesque of Burning Man).[75]

Although ELIZA has been the subject of many, many critical retrospectives, all of which aver the major and enduring influence of the program for the development and current understanding of artificial intelligence, few if any of them are willing to take seriously the subsequent catastrophism of ELIZA's inventor Weizenbaum. Critical approaches to AI generally, however, and to the forms of "machine listening" (still decades away at the time of Weizenbaum's writing), have recently and increasingly operated in the spirit of his writing. They have also stressed the environmental and planetary costs of AI, and they have emphasized the way the white woman's voice synonymous with VUIs like Alexa and Siri obscures the longer, racialized history of domestic service embedded in the history of digital personal assistants, neither of which were explicitly anticipated by the Firesign Theatre or by Weizenbaum's 1970s writing but both of which may be seen as operating in the spirit of their early skepticism, informed as it also was by a curiosity and even enthusiasm about emergent technologies.[76]

In their essay on the Amazon Echo, "Anatomy of AI," Kate Crawford and Vladen Joler have stated that "The 'enclosure' of biodiversity and knowledge is the final step in a series of enclosures that began with the rise of colonialism" and that "the new gold rush in the context of artificial intelligence is to enclose different fields of *human knowing, feeling, and action, in order to capture and privatize those fields*."[77] These are undoubtedly abetted by what Ben Tarnoff has called the "eerily conversational voice" of ChatGPT.[78] If the Firesign Theatre did not in 1971 foresee the possibility of an audience less likely to be horrified by *Bozos'* world of the Future Fair, it might still be consoling to know, at least, that this chapter could never have been written by

```
(PRIN1 (QUOTE "AT LEAST THIS CHAPTER COULD NEVER
HAVE BEEN WRITTEN BY")
    T)
(ADVISE (QUOTE INTERRUPT)
    DOCARM)
(GCTRP 100)
(GCGAG)
;;; STOP
```

5 TELEVISION / What Is Television?

The Martian Space Party (1972)
The Firesign Theatre vs. Dream Monsters from El Outer Space (1972)
Not Insane (1972)
TV or Not TV (1973)
Roller Maidens from Outer Space (1974)
In the Next World, You're on Your Own (1975)

A TRANSIENT AND UNSTABLE MEDIUM

Of all media forms, television is the one most associated with the Firesign Theatre, and it is the one to which they would return most frequently and variously. *Waiting for the Electrician* includes the famous "Beat the Reaper" game show. *How Can You Be in Two Places at Once* begins by parodying low-budget used-car ads produced for late-night TV and then dramatizes the channel surfing that would be explored more fully on *Don't Crush That Dwarf*. Television broadcasts would be woven throughout 1974's *Everything You Know Is Wrong*, and even the Sherlock Holmes parody *The Tale of the Giant Rat of Sumatra* (also 1974) was given the narrative frame of an overnight TV movie in an early draft. On *Bozos*, television is notable in its absence. By focusing on the 1933 Chicago World's Fair, Firesign was avoiding New York's better-known World of Tomorrow Fair, where RCA first demonstrated television in 1939.

The technologies and culture of television changed and expanded dramatically during the years of Firesign's contract with Columbia Records, and as TV increasingly saturated the US, Firesign Theatre albums were routinely understood to perform a kind of television criticism. This was

163

true not only for their satires of what FCC chairman Newton Minow, in 1961, called the "vast wasteland" of programming but also for their attention to the creative possibilities of production and reception. In 1971, the landmark countercultural guide *Guerrilla Television* suggested that the Firesign Theatre, with their use of "special effects, loops, and delays," should be a model for a coming generation of "makers . . . who live together as a collective with videotape, feed back on their experience, and develop technical expertise."[1] And for ordinary viewers, *Dwarf*'s famous use of channel-switching as a compositional device revealed the way the television experience lent itself as a chance-driven "do-it-yourself media collage and even a metaphor for modern life," an everyday practice scholars would be theorizing well into the 1980s.[2] Because remote control devices were still quite rare in 1970—at the turn of the 1980s, they were still only found in 16 percent of US households—Firesign's experiments can be seen as proleptic visions of the medium's future.[3]

But what distinguishes Firesign's work on television is not so much this single discovery but their highly various approaches to television *in the plural*—its programming and its institutions, the many technologies that formed it as a medium, its practices of transmission and reception—all of which were in a significant phase of flux between 1967 and 1975. Taken together, Firesign's television records expressed creatively what media scholar William Uricchio has more recently described as television's status as "a transient and unstable medium, . . . for the speed of its technological change [and] for the process of its cultural transformation, for its ephemeral present, and for its mundane everydayness."[4]

In 1998, the film scholar Charlotte Brundson asked "What Is the 'Television' of Television Studies?"[5] The answer to this question was implicitly multiple. Accordingly, whereas each of the previous chapters has focused on a single album and a single year, this final chapter looks at a range of materials—four albums, a fifth aborted album, a live performance broadcast on the radio that was made into a movie—that variously examine emergent technologies of satellite and cable transmission, the culture of reruns, "ambient" television, and the earliest reality TV.

Each of these projects examined a different television culture or technology, an approach that was intuitive to the Firesign Theatre and perhaps unique in its heterogeny. It was distinct from the other representations of television that proliferated throughout the early 1970s—films like *Brand X* (1970), *The Groove Tube* (1974), and *Kentucky Fried Movie* (1977); the comedy albums of Cheech and Chong; and the direct experiments of *Monty Python's Flying Circus*, which came to US televisions in the summer of 1974. Though popular histories frequently cite Monty Python's censorship battles, it is important to remark the license the BBC afforded Python, not to mention the opportunity itself. The Firesign Theatre never experimented directly with television as Python did, and their albums register that absent possibility.[6]

Timothy Morton and I have described Python's disruptive experiments with television as "an infinite war of meta-escalation. We jump out of the television to find ourselves within the television."[7] (Even *Monty Python and the Holy Grail* concludes with a mass arrest prompted by the accidental killing of a BBC talking head.) The opposite, but complementary, tendency can be found in the Yippies' wish to arrogate the airwaves in service of the revolution, the 1968 Chicago DNC confrontations framed as a form of televised political theater ("the whole world is watching"), which became revised in their lower-stakes appearances on early 1970s talk shows.[8] The Firesign Theatre's instinctive skepticism to this approach, and to the optimistically McLuhanite point in general ("you can't be a revolutionary without a television set"), was forecast in *Waiting for the Electrician*'s fantasy of a turned-on mother imploring her unhip son to "do yourself a favor. Be groovy. Now sit down, turn on, and tune in the TV."

They recorded albums, broadcast live on the radio, worked as Hollywood screenwriters, and would soon publish a book, but for the Firesign Theatre, television was always an affair of remediation. Inspired by William S. Burroughs's "Invisible Generation," they would occasionally tell their radio audience to tune their TVs to channel 11 and turn down the sound while they improvised a new soundtrack. When they returned to KPFK, a recurring bit on *Dear Friends* was to read and discuss ersatz program descriptions from the *TV Guide*:

PB: Let's see what's on. Afternoon, afternoon . . . it's about two o'clock . . . let's see, *End of the World* is on [channel] 2, it's a serial. [Channel] 7 is *The Circumcision Game* and on 4 is *Hopeless Tomorrow*, another one of those serials.

PA: There's a movie on, on channel 9. Drama: *Houston House Detective* (1953). Police Lieutenant Scratch Richards and a urologist named Happy try to track down radioactive chorus girl Honey Sanchez who has a vial of deadly marijuana taped to her leg.

DO: Imagine that.

PB: Who's in it?

PA: Vince Mole, Darlene Edwards, and Elmer Fudd!

PP: He was on last night, he did that special, that *Up America* special, that was a good one.

PA: I think I'd rather watch a movie.

DO: Is there another movie on?

PB: Yeah, *Hercules' Big Arms* is on. That's one of those 60s Italian macho films I think: "Hercules tries to keep his mind on avenging his dead wife." I saw it, I saw it! Mickey Mulky, Benito Mussolini. It's 90 minutes, and it's general public . . .

PA: No, no, better you'd like Movie (science fiction) on channel 13 *Space Delinquent* (1957).

PB: Great year man, great year.

PA: Listen to this: A teenage Martian tries to sell some hot marijuana to a federal agent when there is a strange noise from space. Helen Whipple, Maurice Bumstead.

PP: [whistles like a space ship]

PB: Oh yeah hey listen.

DO: The afternoon is really a bad time to watch TV, guys.

PB: No, no, there's some of that NET stuff, some of that free television, you know. Walt Disney, *Dances of Cheese*: Traditional cheese dancing is on 28.

DO: That's a bore! . . .

PA: Here's a new game, a debut it says *J'accuse!*

PB: Oh yeah?

PA: From France comes this popular favorite in which the people (Jan Murray) tries to kill all the guest aristocrats without dropping a stitch of laughter. This week's guests: Barbara Bobo, Adam Thigh, Willie Pep, and Whittaker Chambers.

PB: Whittaker's getting back!

DO: Swords will be sharpened for that.

WHAT IS TELEVISION? 167

Written in advance by Phil Austin, and riffed on by the group, these were a favorite of fans, too, as can be seen by a page of the *Chromium Switch* fanzine, parodying *TV Guide* with numerous Firesign citations (fig. 24).

The Firesign Theatre did not originate critical television parody, of course. Besides the influential commercial television satires in *Mad* magazine (est. 1952), Phil Proctor has often spoken of his admiration for Ernie Kovacs, whose 1950s television shows "specialized in absurd visual tricks, elaborate set pieces, and anti-realist montage symphonies that juxtaposed rapidly edited and incongruous images against music . . . us[ing] sound counterintuitively and sometimes with no particular relation to the image at all."[9]

By the turn of the 1970s, the general attitude to television shared by the Firesign and their audience—disdain leavened by curiosity—was emblematized by a reference on the first, self-titled album by Cheech and Chong (1971). Having already quoted Firesign directly ("long in the leaf and short in the can"), the album closes with a skit involving aimless stoned channel surfing, in which it is discovered that the TV needs to be turned on. Eventually, they abandon the TV and play an album on the stereo instead. Unfortunately for Firesign (and this is a moral as well as a punch line), the album turns out to be the first Cheech and Chong record.

Defining themselves in early interviews as identity-based comedy (as opposed to what they saw as the implicitly white college-oriented comedy of Firesign), the Mexican American Cheech Marin and Asian Canadian Tommy Chong would have a commercial success that far surpassed that of the Firesign Theatre, culminating in a series of major studio movies in the late 1970s. In retrospect, their ambiguous homage is an emblem of Firesign's mutable fortunes in the years covered in this chapter, years that extend from the peak of their popularity to the termination of their Columbia contract in 1975.

This did not seem inevitable in early 1972. A ten-paragraph feature in the *New York Times* in February preceded Barbara Garson's *Harpers* article in June, in which the activist playwright examined labor unrest from a post at an "auto-workers commune" near Lordstown, Ohio. There, young people "fed up with the whole industrial system" could be heard listening together to the Firesign Theatre, suggesting the degree to which Firesign's audience extended beyond the college audience.[10]

CHROMIUM SWITCH.....page 3

3:12 **4 OZZIE KNOWS FATHER** BW
  6 WHO DO YOU TRUST?—Game BW
  Guest: John Ehrlichman
3:30 **2 MUTT 'N' SMUTT**
  Miss Doody finds her daughter buying a thermal throw
  robe and some adjustable toe regulators from the
  junk man Mutt.
  7 DR. BEANBAG
  26 THE LES CRANE SHOW BW
  Phil, Phil, Phil, and Phil of the Fireside Theatre never
  reveal themselves in this revealing hour. (45 min.)
4:01 **2 CHANNEL CITY NEWS**
  10 DR. WHIPLASH
  Your host: Phil O'Steen
4:06 **10 BITING THROUGH—Travel**
  Holy-graphic journeys with Clem/George and guest
  announcer Floyd Damme. (25 min.)
4:30 **10 CHEESE-BOMB SOCCER**
  Mixville vs. Smoggywood (Live; 10 min.)
4:57 **65 SIGN-ON/FRED FLAMM NEWS**
5:00 **4 FREAK FOR A WEEK**
  26 HOWL OF THE WOLF MOVIE BW
  "Highschool Madness." (1954) Dave Casman, Joe
  Bertman. (57 min.)
5:19 **8 MOVIE—Drama**
  "Mud-Wrestlers from Outer Space vs. The Mexican
  Mole Women." (1964) A nuclear physicist discovers
  that he has strange desires. Gunther Montez, Richard
  Basehart. (2 hrs.)
  13 MOVIE—Sci Fi BW
  "Space Delinquent." (1957) A teenage Martian tries to
  sell some hot marijuana to a federal agent when there
  is a strange noise from space. Helen Whipple, Maurice
  Bumstead. (1 hr., 20 sec.)
6:00 **7 HAWAIIAN SELL-OUT—Game**
6:18 **10 CRAZY AS LOONS—Phil Yamamoto**
  65 CIRQUE INTERNATIONALE
7:06 **65 EMERGING FALL OF THE ROAMING UMPIRE**
  Your host: Freddy Burns
7:45 **8 EVERY NIGHT AT THE MOVIES** BW
  "Mother and Child." Lilly Lamont, Erpo Sweeny.
  14 LIVE FROM PANORAMALAND 2000
  Historic replay of Time and his escape from the "Gilda."

THE ROXY THEATRE (IN TORONTO) RAN A 26-HOUR
NON-STOP SCI-FI AND HORROR MOVIE FESTIVAL THIS
SUMMER. PLACED BETWEEN "BARBARELLA" AND
"NIGHT OF THE LIVING DEAD" WAS NONE OTHER THAN
"MARTIAN SPACE PARTY."

Figure 24. "T. B. Glide" in Tom Gedwillo's fanzine *Chromium Switch*, October 1973. Firesign Theatre Collection, National Audio-Visual Conservation Center, Library of Congress.

The independent rock press, meanwhile, was increasingly treating the group as being so preeminent in the expanding comedy field as to have transcended it entirely, which would prove to be a precarious distinction. In a roundup of the "New Comedy" in *Rock* magazine, the Conception Corporation's Murphy Dunne was quoted as saying, "The Firesign Theatre state precisely that satire and to be funny is not what they do. *They're more surrealists, creating an entire new environment.*" The article added that this was why Firesign was not represented in the piece and promised a stand-alone Firesign feature soon.[11] That article did not materialize. Featuring DJ-friendly selections from their radio shows, Columbia had released the double LP *Dear Friends* compilation in February, promoting Firesign as "the only rock band in the world that doesn't need music." But that advertisement appeared in the context of an FM radio that was becoming steadily less hospitable to the Firesign Theatre (as discussed in chapter 2).

The entirety of the group's catalog was nevertheless selling well in the spring of 1972, inspiring Columbia to issue a second five-year contract and greenlight a project titled *The Martian Space Party*—a live perform-ance at the new KPFK building that would be captured on sixteen-track tape (via a remote truck) and shot as a movie on three 35 mm cameras while also being broadcast live on the California Pacifica FM stations.[12] By then, Firesign had signed arrangements with *Rolling Stone*'s Straight Arrow Press to publish a book of the early albums' scripts, which would inspire community theater productions that lasted throughout the dec-ade. A further intriguing development in this 1971–72 period is the increasing role of Annalee Austin and Tiny Ossman. Both, especially Annalee, had more prominent roles on *Bozos*, were pictured on the back cover, and would be featured still more prominently in *The Martian Space Party*. At a 1972 photoshoot, all six appeared, as if the band had expanded.

At the same time, there were changes in the group members' lives, ten-sions within the group, general fatigue, transformations in the field of popular culture, political conflicts, and setbacks—all of which weighed on them. It had still not been long after the deaths of Bergman's girlfriend Linda Sugarman and dear friend Jere Brian (the latter scuttling plans for a Firesign film), and these were soon followed by the death of Proctor's friend Brandon de Wilde in a traffic accident in July. The Ossmans had moved to Santa Barbara after recording *Bozos,* which substantially affected the group's working rhythm, even though David continued to spend three to four days a week in LA. Finally, there was an ambient sense that the group should be still more successful; they were at last making a low-budget movie, but the first of the period's rock operas were now going into production with the full weight of the film studios behind them. Austin in particular was frustrated by the group not having the success of a rock band, but he also refused to fly, which made touring difficult.

Starting with the Rolling Stones' 1972 US tour (their first since Altamont), touring became reconfigured on an industrial scale, leading to the following year's arena and stadium tours by Pink Floyd, Alice Cooper, Led Zeppelin, and the Who. The profits raked in by these megatours ameliorated the surging price of vinyl, which had skyrocketed as a result of the 1973 energy crisis.[13] And they were sustained by technological developments in PA systems, which had improved dramatically after Woodstock in August 1969, transforming the scale and the experience of live shows.[14] In 1973, Pink Floyd's *Dark Side of the Moon* and The Who's *Quadrophenia* tours featured 360-degree quadraphonic sound systems and twenty-eight-channel mixing boards that could also cue in backing tapes of ambient sounds like clocks, waves, train whistles, and prerecorded musical parts.[15]

Firesign's earliest live performances had been at a small club called the Ash Grove, appearing with musical acts like the Byrds, Buffalo Springfield, and Taj Mahal. For the KPFK "Sound '68" benefit at the Santa Monica Civic Center, they shared the bill with the Springfield, Blue Cheer, Dick Gregory, Sonny Terry and Brownie McGhee, and the United States of America. Their spring 1970 East Coast tour took them mostly to college theaters; around this time, they were also offered a spot opening for the Beach Boys on tour, which they declined. On their 1974 tour, they played at a mixture of universities and rock venues (the Celebrity Theatre in Phoenix and Kansas City's Cowtown Ballroom), crowned by a sold-out show at Carnegie Hall. According to *Chromium Switch,* they were selling between fifteen hundred and twenty-five hundred tickets for these shows, but the scale for the premiere rock acts had by then transformed entirely.[16] A month after Firesign played Carnegie Hall, the Who sold out four nights at the twenty-thousand-seat Madison Square Garden.[17]

And there were issues related to the work itself. On *Bozos*, the Firesign Theatre had wishfully killed—or, if you prefer, *broken*—the president. But as 1972 rolled around, the Vietnam War was still raging, with Nixon enjoying his highest approval rating in two years (61 percent), well on his way to crushing George McGovern in the election that fall. In his journal on April 12, Ossman writes that he was "totally distracted by The War on TV—shit!" As Firesign prepared to enter the studio, Ossman notes the group's depression when at the end of the day, Nixon announced on television that the US would mine North Vietnam's Haiphong Harbor.[18] He was withdrawing troops but intensifying bombing. Phil Proctor would remember that "we knew it was the end of the revolutionary period for our country."[19]

"What can you say about a cultural revolution that died?" wrote Rolling Stones fan Ellen Willis in her review of the Rolling Stones' four Madison Square Garden shows that July. She would later write that by the spring of 1972, "all over the country groups of people who tried to live by an ideology of leaderlessness were disintegrating in bitterness and confusion."[20] It turns out that those words applied very well to the Firesign Theatre, which had always arrived at decisions by consensus. In June, just after the *Martian Space Party*, the Firesign Theatre split, from pressure, exhaustion, and, perhaps above all, the collapse of Phil's marriage to Annalee—which entailed an affair with Bergman and a short-lived plan for her to replace Austin in the group. In October 1972, just as *Creem* published the first of two huge features on the Firesign Theatre, *Rolling Stone* announced that the group had broken up.[21] This was followed by vague retractions, but the Firesign Theatre would not work together again until August 1973, and by then, things were very different. One casualty of 1972 was *The Firesign Theatre vs. Dream Monsters from El Outer Space*, a hugely ambitious fifth album that was to have grown out of *The Martian Space Party*. Should we watch some TV?

SATELLITE

In November 1971, Firesign returned to KPFK for a new series of broadcasts called *Let's Eat*, which, like *Dear Friends*, would be offered to commercial FM stations and closed-circuit college stations. With *The Martian*

Space Party planned as its final broadcast, *Let's Eat* was meant to be a laboratory for Firesign's next studio album, as had been the practice with the previous records.

At first, it seemed that the new record would develop the group's clever but somewhat tiresome Shakespeare parody, which they had performed as *The Count of Monte Cristo* and now bore the title *Anything You Want To*. As spring wore on, a more ambitious plan emerged, inspired by such contemporary events as the Apollo 16 lunar mission, the ongoing Vietnam War, the Sapporo Winter Olympics, the looming presidential election, and Nixon's historic visit to China, which would produce a presidential meeting with Mao Zedong and the first broadcast images of China in twenty years. The Shakespeare material was retained but was no longer the guiding principle; all the other contexts were the domain of satellite television.

Though it is more than a little shaggy, *The Martian Space Party* is framed as a television broadcast monitoring two major concurrent live events.[22] The first of these is the presidential nominating convention of the Nat'l Surrealist Light Peoples Party, which is taking place in "the wonderful resort town of San Clamaron on beautiful Gas War Island." With the KPFK performance space decorated with streamers and balloons, the audience were addressed as delegates embraced by the Surrealist Party's big tent: the National Association of Funny Name Clubs of America, the National Spider Caucus, the League of Wingéd Voters, the March Eighth Laughing Words Woman's Shock Team, the More Sugar Foundation, and the American Friends of the Martian Space Party. After several speeches, interviews, and songs—the latter performed by the four Firesigns, Tiny, and Annalee (fig. 25)—the storyline culminates in the unanimous nomination of George Papoon for president: "You know he's Not Insane!"

Coverage of the Surrealist convention is continually interrupted by bulletins from the other major news event, the current president's state trip to Monster Island. This at first consists of reports of the president's repeatedly being refused entrance to the Forbidden City, which causes the "emotional Christian man" to cry. Presenting and commenting on these events, the four Firesigns assume the personae of respected American news broadcasters: Proctor and Austin jointly perform the role of the home studio's "Walter" (Cronkite); Ossman, as "Eric" (Sevareid), and Bergman

Figure 25. The Firesign Theatre's Martian Space Party, KPFK performance space, March 30, 1972 (*left to right*, Annalee Austin, Tiny Ossman, Phil Austin, Peter Bergman, David Ossman, Philip Proctor). Photo by Jerry De Wilde. Permission of the artist.

(*Waiting for the Electrician*'s Charles B. Smith) communicate by satellite from Monster Island:

ERIC: There's so much terrible noise here that it's made the president cry.

WALTER: Tell me again about the monsters, Eric.

ERIC: Well it's not so much the monsters here, Walter, as the dreadful *fear* and the senseless destruction that they inspire in *all* of us. Even the president has been deeply affected by them. Anyone can see that by the fact that he just doesn't know what to do with his hands.

Their performance mimics the technical difficulties common to live broadcasts (randomly cutting off words in the middle of a sentence), as well as those that might plague its preprogrammed events (unpredictably speeding up to the point of unintelligibility, in emulation of malfunctioning videotape). In this way, they are sending up what television scholar

Lisa Parks has called the "fantasy of global presence" that characterized satellite broadcasts such as the multinational spectacular *Our World* of June 1967 (where the Beatles had live premiered "All You Need Is Love"), as well as dramatizing its material conditions.[23]

But it also dramatized the political force of television generally, and the evolving meaning of *satellite* in particular, five years after the techno-utopianism of *Our World*. This was emphasized by the allusions to Sevareid and Cronkite. Both had been critical of the Vietnam War, with Cronkite's famous February 1968 broadcast from Saigon, after the Tet Offensive, widely recognized as a watershed moment: "the bloody experience of Vietnam is to end in a stalemate." Cronkite's apostasy had been given authority and enabled by the new technologies of videotape and direct satellite transmission, which all three commercial networks had just begun using to relay footage from Japan to New York. Because of them, the Pentagon was no longer controlling the public image of the war.[24]

In August, the televised chaos of the Chicago DNC was also broadcast live. But by 1972, there was a plangent sense that political culture was once again subject to informational control and domination. That was nowhere clearer than in the satellite-enabled "television diplomacy" of Nixon's February 1972 visit to China, which sent images of "spectacular banquets, toasts, handshakes, and smiles, but . . . nothing of what was said in off-camera talks, beyond a report that the two powers had agreed to 'normalize' relations, and to trade."[25] On the journey back, Air Force One spent hours on the Alaska tarmac so that Nixon could finally arrive in Washington during prime time (in an election year). The fact that Firesign was acutely aware of these politics is confirmed by the single page of the May 19–26 *TV Guide* preserved in Phil Proctor's archive: it notes in advance that Tuesday's 6:55 *Political Talk* with candidate George McGovern "may be pre-empted with special reports" "if President Nixon's trip [to Russia] is proceeding as planned."[26] The coverage of the president's frustrated trip to the Forbidden City on Monster Island was an attempt to name and deflate these tactics, while the Papoon convention, in contrast to the memory of Chicago and reflecting the new sense that the revolution would not be televised, makes a joke of its own boredom and inanity.

This vision of the resulting meaning of satellite television accords with Raymond Williams's prognosis two years later. Appreciating satellite's potential, Williams admits that a McLuhanite "world-wide television service, with genuinely open skies, would be an enormous gain to the peoples of the world." But "most of the inhabitants of the 'global village' would [most likely] be saying nothing, in these new terms, while a few powerful corporations and governments, and the people they could hire, would speak in ways never before known to most of the peoples of the world."[27] As Parks stresses, the 1960s fantasy of a satellite "global presence" had begun to give way to a reality of "global monitoring," and Firesign's comprehension of this transition would become clearer when they began to revise *The Martian Space Party* as an album.[28]

While these synchronous news broadcasts formed the center of the *Space Party* performance, the television fiction extended to other stations on the dial, where we get snippets of the reworked Shakespeare parody, more advertisements, *The Hilerio Spacepipe Show*, and *Young Guy: Motor Detective*, a "Japanese" rewrite of Nick Danger that paid homage to the Japanese movies and television programs that had begun to appear on US television in English-language dubs by the late 1960s. Inevitably, though, *The Martian Space Party* culminates in the interlinking of the two live broadcasts. As George Papoon secures the Surrealist nomination, the Nixonian president, having met with the residents of Monster Island, prepares to blast off from the island on a rocket for Mars, which is at the last minute hijacked by the ancient seventeen-foot monster Glutamoto, who unfurls a banner from his tail reading "Not Insane!"

This allegory of Nixon's visit to China merits critical analysis, both for its elision of Vietnam (which was Nixon's point, too) and for its dubious conflation of Japan and China. Glutamoto was transparently a version of Godzilla, of course, and the island location itself was borrowed directly from Toho Studios' late-1960s monster films: *Son of Godzilla* (1967, US version 1969), *Destroy All Monsters* (1968, US version 1969), and *All Monsters Attack* (1969, US version 1971) all take place on Monster Island.[29] In addition, the *Ultraman* television program—a dubbed version arriving on US screens in 1972—was named in Firesign's script drafts, eventually becoming Young Guy's atomic butler Rotonoto; Young Guy was himself a citation of the title character of the Wakadaishō comedy-action films, which featured actors from the Toho monster movies. These were realized in "aural yellowface" performances that were transparent expressions of both love and theft.[30]

Different from McLuhan's racist hailing of the "Oriental" sensibility supposedly heralded by the age of electronic communications, Firesign's orientalism was produced in the context of, and despite, numerous long-established Asian communities in Southern California. This cultural diversity was more visible now on Los Angeles television, which no doubt made it more available as a source in the otherwise deeply segregated city. By 1972, there were two UHF stations broadcasting television in Spanish, KMEX (34) and KLXA (40), as well as KWHY (22), which broadcast programming in Spanish, Korean, Japanese, and Chinese.[31] And there was increasing programming by and for Black audiences that included PBS's *Soul!* (1968–73), *Soul Train* (which debuted in 1970), and Cincinnati's *Soul Street* (1972).[32] It is surely no accident that this would also be the period when Firesign employed the most, and the most varied, racialized voices. They expanded the range of their heteroglossia, but they did so at a time when other popular recordings, by actual comedians of color like Cheech and Chong and Richard Pryor, were increasingly underscoring them as problematic, showing Firesign to have been a hitherto unremarked performance of white identity whose domain was everywhere and nowhere.[33] Without abandoning their political commitments, the Firesign Theatre's last Columbia albums, which were pointedly not set in LA, were ones on which they reconsidered the meaning of these vocal techniques.

After the March 30 *Martian Space Party*, the Firesign Theatre spent the next month editing the hours of footage into a thirty-minute film. In May, they turned to the question of the new album, which they provisionally titled *Not Insane, or The Firesign Theatre vs. Dream Monsters from El Outer Space*. The group wrote extensively for the next two weeks and began recording at Columbia Square on May 15. Nine days later, the sessions ceased abruptly, and the album was never completed.

What survives of the *Dream Monsters* project is sparse but very suggestive. Acknowledging that the resulting album would have changed, given their iterative practice of recording and writing, it is still worth attempting a speculative reconstruction. In addition to *The Martian Space Party*, nine minutes of recording survived, appearing eventually on the *Shoes for Industry* compilation in 1993. In addition to those recordings are seventeen pages of script, two pages of plot outline, and some figural sketches.

All these fragments suggest that the *Dream Monsters* would have turned away from the simple linearity of *Bozos*, returning to (if not exceeding) the structural and sonic complexity of *How Can You Be* and *Dwarf*. And it is clear that this complexity would have had everything to do with satellites.

This is indicated by the conceptual drawing Ossman made on the back of the provisional outline of the plot (fig. 26). At the top of the page is Mars, the ultimate destination of the *Martian Space Party*'s president. Beneath it is the moon, and beneath the moon are two satellites which were to have been personified, "as if programs themselves and their creators were orbiting around the earth, looking down at the Earth as they did their shows."[34] One of these satellites is the province of an oily talk-show host named Hilerio Spacepipe (a riff on Hugh Hefner's short-lived CBS program *Playboy after Dark*). The other is a revision of the humane correspondent Walter, who is now mechanized as the ominous W.A.L.T.E.R., the Watch And Listen To Everything Robot. This was a recognition that the same satellite technology that had transformed the meaning of the Vietnam War (as everyone agreed) would also be instrumental in its "remote sensing" capacity for military control and monitoring.[35] Beneath the satellites, and separated by the Great Wall, are Monster Island and Gas War Island, the sites of *The Martian Space Party*'s fictional broadcasts. And at the bottom are the president and several characters from *Young Guy*.

The script drafts suggest that this atmospheric geography would be traversed by two additional storylines. In one, Ossman's campy 1950s spaceman hero Mark Time is returning from Planet X in a stolen bread truck called the Mixville Rocket, passing through layered waves of satellite transmission while communicating with WALTER ("This is Walter: you have entered Earth-Wake but communication is dubious. I will attempt to race the shadow and revise you"). Beneath the layers of transmission is "Radio Prison," which is where we find Phil Austin's Young Guy listening to a pirate broadcast of a DJ called The Lizard. "These stories," Ossman's diary notes, "would intersect in a virtual world of merging media, watched over by WALTER—the supreme Watch And Listen To Everything Robot."[36]

The script further indicates the major aural and narrative role that would be given to the satellite broadcasts, describing eight distinct tracks

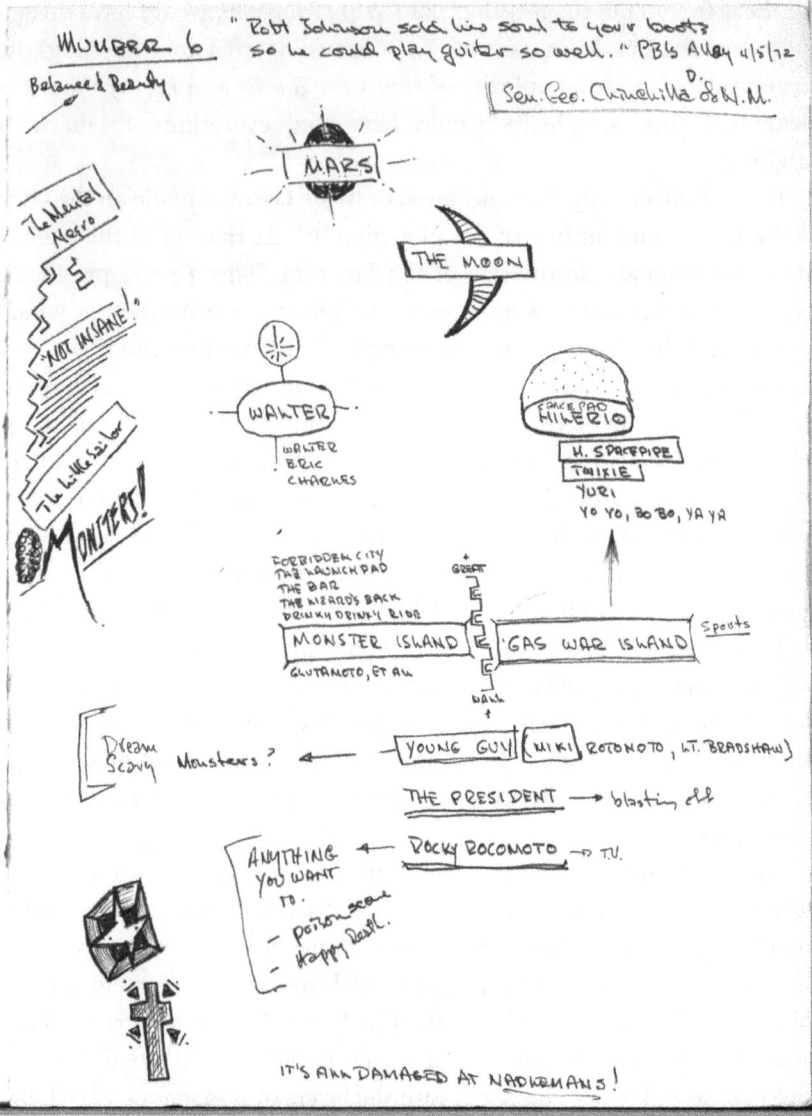

Figure 26. David Ossman's draft materials from *Not Insane, or Dream Monsters from El Outer Space*. Firesign Theatre Collection, National Audio-Visual Conservation Center, Library of Congress.

Figure 27. Papoon victory announced on the front page of the *Michigan Daily*, April 1, 1973. Courtesy of University of Michigan, Office of Student Publications.

that were to have been recorded—and mixed in and out of the narrative—to resemble "vintage radio, foreign broadcasts and TV movies, to the noise of solar wind, all created from recycled ads, readings of Filipino comic books, gospel music, coverage of the Olympics in Tierra del Fuego and other bits from the *Let's Eat* radio shows."[37] These broadcasts, which can be briefly heard in the *Shoes for Industry* mix of "Young Guy," are a dramatic representation of what Hito Steyerl has much more recently termed "spam of the earth": "Dense clusters of radio waves leave our planet every second. Our letters and snapshots, intimate and official communications, TV broadcasts and text messages drift away from earth in rings, a tectonic architecture of the desires and fears of our times. In a few hundred thousand years, extraterrestrial forms of intelligence may incredulously sift through our wireless communications."[38] *Dream Monsters* would be a story that took place in outer space, but it was not the silent empty space of Kubrick's *2001*; rather, it was a grievously noisy space populated by electronic human noise: "How am I to be Not Insane," groans Young Guy, "while suffering [the] terrific torment of Radio Prison?"

With the group in disarray, it fell to Austin and Proctor to compile an album from what remained. As released in October 1972, *Not Insane* was composed largely of a two-year-old performance of the Shakespeare piece together with excerpts from *The Martian Space Party*, culminating in the celebratory nomination of Papoon. Obviously fragmentary and extravagantly treated with flanging effects, it is somewhat amazing but deeply frustrating to listen to and does not elicit many further discoveries on a second hearing. This was confirmed by separate reviews in consecutive issues of *Creem*, the first saying it was "so thick and heavy that it is not only not very funny, but downright hard to listen to unless you're totally zonked," the second much longer review sadly admitting that "while this LP is funny, it is not profound."[39] (Its entry in *Motorbooty*'s 1992 survey reads: "shit sandwich.") Unexpectedly, however, the Papoon—Not Insane campaign rapidly became one of Firesign's most omnipresent memes. On April 1, the headline of the *Michigan Daily* read, "Papoon Takes City in Mudslide" (fig. 27); that same day, John Lennon wore his Not Insane pin to the Nutopia press conference. And as we will see, the Papoon concept would organize an enormous amount of the fans' collective work later in the decade, as Firesign itself became more obscure.

CABLE

After the collapse of *Dream Monsters from El Outer Space*, the Firesign Theatre would next come together for the big screen debut of *The Martian Space Party* at the Director's Guild Theatre on June 29. Released as a package with the recently rediscovered *Reefer Madness* (1936), the fledgling New Line Cinema circulated the film throughout the fall of 1972.[40] Columbia Records gamely released *Not Insane* in October and then, apparently unfazed by the group's status, greenlit solo records for each of the Firesigns.

This was also the semester Raymond Williams spent as a visiting professor of Political Science at Stanford, where, sitting "in our flat in Escondido Village and often watch[ing] American television almost around the clock," he drafted the seminal book *Television: Technology and Cultural Form*. That book is most famous for Williams's concept of *flow*, an idea that was not originally intended to describe the creative experience of channel surfing. As was also true for the Firesign Theatre, Williams is there concerned with much more than the phenomenology of viewing and extends his sense of television's *users* to include broadcasters and institutions. They are the authors of "flow," in the term's original conception.

"The other useful thing which happened at Stanford," Williams would remember, "was contact through its Department of Communications with the newly emerging technologies of satellite and cable transmission."[41] As noted above, Williams appreciated the promise of satellite television and feared its likely cooptation. He had similarly ambivalent words about the prospects for cable television, which was at a crucial stage of development in 1972 and 1973. The first of the Firesign Theatre's solo albums, Proctor and Bergman's *TV or Not TV* (1973), can be understood as an expression of this same ambivalence. *TV or Not TV* was also, more directly, representative of the two Firesigns' desire to make records that were more accessible and to tour (which resulted in an invitation to meet Marshall McLuhan in his chambers at the University of Toronto, where he gave them exploding cigars). But, centered as it is on two characters who are running a subscription cable service in 1973, it also engaged frankly with the contradictory currents of a medium in a key period of transition.

Cable—or, as it was originally named, Community Antenna Television (CATV)—had a history in the US that dated to the 1940s, when it was used to reach viewers living in rural areas beyond the reach of conventional VHF and UHF broadcast signals. The established broadcast networks, alive to the potential challenge to their monopoly, then lobbied the FCC for regulations that would ban CATV services from the one hundred biggest cities in the US and prohibited their use for any programming other than the retransmission of local broadcast signals.[42]

This arrangement, and even the conception of the technology's possibilities and implications, began to undergo a dramatic reversal in the late 1960s, a change emblematized by CATV's rebranding as "cable television."[43] This entailed a confluence of changed attitudes among parties that might otherwise have been expected to be in conflict. There was a liberal consensus advocating for the creation of cable channels for "educational, government, and public access," emphasizing the transformative potential of locally produced programming as a social good.[44] This understanding was taken to have revolutionary potential by the Yippies, whose 1968 Chicago manifesto included the demand for "the free and open use of the media. A program which actively supports and promotes cable television as a method of increasing the selection of channels to the viewer."[45] Sympathetically aligned with the Yippies was *Guerrilla Television*'s "Official Manual," which offered advice on how to educate cable television programmers in preserving the values of cable as a narrowcasting medium.

These ideas were notably, if ironically, aligned with what became the dominant understanding of cable among policy makers: a "Blue Sky" discourse that envisioned cable as "an electronic highway" and "the television of abundance," the latter a phrase that originated in an influential 1971 Sloan Foundation report, which argued that cable would bring to television a "particularity" and "scope" that would make television "commensurate with the press," a crucial component of a democratic public sphere.[46] As communication scholar Megan Mullen has written, between 1968 and 1974, "policymakers tried to force a complete reversal of cable's primary function: from a basic rural retransmission medium to the locally oriented and content-specialized medium broadcast television could never be. Local programming strategies were developed for cable. The concept of public access programming was introduced. The earliest pay-cable networks were started."[47] At the same time, however, cable operators also won the ability to import distant signals (including, eventually, from satellite). In 1972, the FCC permitted cable distribution in the top one hundred US markets, the "watershed moment" when cable evolved from being a threat to the broadcast industries to being a component within it, a process that was completed by 1975.[48] Reviewing this history in 1997, the sociologist Thomas Streeter observed a similar process unfolding among the techno-utopian evangelists of the early Internet.[49]

As Williams would later write, "the changes were politically willed and managed," which was also to say that it was not the "essential and inevitable destiny of the medium."[50] In the early 1970s, cable was still understood to have democratic potential, including that of a two-way interactive technology.[51] Williams, for instance, recognized the possibilities for education, stressing that public-access systems, "genuinely run by and serving local communities," "could indeed democratise broadcasting." But he also could already identify a cable industry whose "best financed ... companies," driven by advertising profits, "are offering what is, essentially, a version of the very worst kind of broadcasting service."[52] To consider the culture of cable television at that moment—as Proctor and Bergman did when they entered the Wally Heider Studio in LA for *TV or Not TV*—is to encounter it in a period of undecided possibilities.

Proctor and Bergman's investigation of cable television expresses less of Williams's residual optimism. *TV or Not TV* does not concern a public access channel but is instead structured as a broadcast day on a small subscription cable service, channel 85's "Scramble Vision, a Free Service of Charge-a-Card," run from the home of one of its two operators, the comically milquetoast Fred Flamm (Proctor) and Clark Cable (Bergman). The day's programming is entirely composed of prerecorded material, accessed by Fred and Clark from a library of videocassettes, and the record begins by emphasizing how the broadcast day is almost wholly devoid of the liveness that defined the fiction of *The Martian Space Party*. Fred Flamm addresses the audience at the beginning by saying, "Good afternoon, morning, or good evening" before Clark comes in to add, "It's my job to keep things lookin' real, even if we are on tape," which leads to confusion as to whether they are in fact "now" on tape.

The broadcast day consists of three more or less conventional entertainment programs: a talk show, a broadcast of a circus performance, and a parody of PBS's *Masterpiece Theatre* (which had premiered in 1971), "made possible by a begrudged grant from USUK . . . a short division of Burns Psychotic Plastic." These are strung together with periodic reminders to insert a credit card for continued viewing, some public-service announcements, and an "opinion to the contrary" from Citizen Milton Spurt, president of Spurt Furniture, who opines, "If I'm going to be charged in a free society, then it is my view that opinions expressed, editorial or otherwise, should be carefully eliminated."

To the extent that there is a narrative beyond the programming, it consists in the interruptions of a group of teenagers who can be heard unpredictably hacking into the channel 85 transmission. The hacks are more puerile than political, however. Under the banner of "Youth Wants It Now," they demand, almost inaudibly: "1. No more school! 2. Give us our vote! 3. Dump Children's Commissioner Mertz! 4. Abolish statutory rape for consenting teenagers!" This is as distant an echo of the Yippies' "tactical manipulation of television" as the prerecorded programming is very far from the disruptive work heralded in *Guerrilla Television*. With that book,

Shamberg hoped to galvanize a generation of video collectives, using the new PortaPak video recorder in the service of a reflexive media criticism, inspired by direct cinema and the recursive production effects heard on the earlier Firesign Theatre albums. Although channel 85 seems on one level to be a salutary form of narrowcasting, the broadcast itself appears to be an almost programmatic rejection of these countercultural tactics.

An open question is whether *TV or Not TV* means to reject the radical television movement as such—does the album title mean to cite Shamberg's San Francisco collective TVTV (Top Value Television)?—or if it instead ruefully mocks the aspirations of cable utopianists. It is indeed not hard to think that channel 85 is "offering what is, essentially, a version of the very worst kind of broadcasting service." That Proctor and Bergman seem to be saying something similar seems intimated by the doublespeak of the programmers, who repeatedly remind us of the free services that we have to pay for.

This is represented in another way by the longest sustained fiction on the album, the *Masterpiece Theatre* parody titled *The Declining Fall of the Roaming Umpire*. Featuring astonishing performances (Proctor in three roles, Bergman in one) and breakneck wordplay, the episodic drama slowly reveals itself as a parable on the problem and promise of cable television, circa 1973. Set in ancient Rome ("Roaming"), the first installment hinges on a conversation between two senators, Flattus Prolongus and Mucus Bruto. Bruto has had a dream he believes is an omen, and Flattus encourages him to "tell a vision." We recognize that Bruto's premonitory dream was about a baseball game between the Senators and the Angels at "La Coliseum." Bruto is shocked when he sees an Umperor eject a Senator with his thumb and shocked further when the crowd cheers in approval. "But there was a hopeful sign. The crowd did not cry, as here in Roaming, 'Long live the Umperor,' but rather '*kill him.*'" Bruto and Flattus take this as inspiration to kill the emperor Caliuga.

So far, this skit relates to television only insofar as Bruto's dream is maybe a television broadcast sent from the future, though it may also (via the pun on Washington's baseball team the Senators) be a dim metaphor for deregulation. When the narrative resumes, however, it becomes explicitly an allegory about cable television. In this case, it is the Umperor Caliuga (his name conflating the Roman Caligula with the cataclysmic world age of Hinduism, Kali Yuga) who has had a portentous dream:

CALIUGA: I see a tower—what an eyeful! I'm going to build a tower of power, and I shall rise up and direct the building of great *networks*, and broad cast them upon the waves— one voice, one image, one Umperor, one me!

FLATTUS (ENTERS): And one is nothing!

CALIUGA: What? Who gives me such poor reception? . . .

FLATTUS AND BRUTO: A congress of senators!

CALIUGA: A couple of eggheads! Come to scramble my vision!

FLATTUS: Exactly: you've poached the people's power.

BRUTO: And there's no gold left to lay another egg.

CALIUGA: I'll lay anything I want! All hail Caliuga, I shall reign forever!

At this point Caliuga is assassinated, an apparent allegory of the triumphant Blue Sky narrative of the egghead technocrats, wresting control from the authoritarian monopoly of the networks and fostering the hundred blooming flowers such as channel 85. As Flattus self-assuredly proclaims, "He's dead, the show's over, long live the people! . . ."

But the meaning of the piece then changes completely as Flattus adds, "And now, at last, *the wealthy and the powerful can speak for all*!" Spoken here by Bergman, these lines prove a canny analysis or prognostication of what was even then happening, as the FCC's 1972 deregulatory decision "transformed cable from a community service into a business."[53] And this was all operative at another allegorical level, since the historical Caligula had been assassinated in a plot that hoped to reestablish the Roman Republic. That hope was suppressed when the Praetorian Guard instead installed Caligula's uncle, Claudius, as emperor. Imagine Proctor and Bergman's amusement when *Masterpiece Theatre* broadcast the BBC's twelve-part adaptation of Robert Graves's *I, Claudius* five years later.

RERUNS, DREAMS

A month after Proctor and Bergman recorded *TV or Not TV*, David Ossman booked Wally Heider Studio for his album *How Time Flys*, a radio drama distantly related to the "Mark Time" story drafted for *Dream*

Monsters from El Outer Space. The cast included David and Tiny Ossman, Harry Shearer, Wolfman Jack, the Credibility Gap's Lew Irwin, the three other Firesigns, and Proctor's wife, Sheilah Wells. Though the cover bills it as a "hi-fi sci-fi comedy," *How Time Flys* is far straighter than the other Firesign albums. It is also only tangentially about television: Mark Time is returning to Earth with a hologram recording of his journey to the Planet X; he hopes to provide the recordings to the public via the news media, while the devious Mr. Motion wants to steal them and sell them as entertainment.

A year later, Phil Austin's *Roller Maidens from Outer Space* was released on the CBS subsidiary Epic. By that time, Clive Davis had been fired from Columbia, *Rolling Stone* had been reformatted as a general-interest lifestyle magazine, and, as Ossman put it, "the accountants had taken over the business."[54] If *TV or Not TV* does not definitively belong to this period, Austin's album and the subsequent Columbia Firesign Theatre albums do.

Named by Chicago punk band the Dwarves' Blag Dahlia in 2019 as his favorite album of all time, *Roller Maidens* features Austin in multiple acting roles and fronting a band that included Cyrus Faryar and Henry Diltz (both formerly of the Modern Folk Quartet), as well as Austin's new partner (Oona Elliott) and Lesley Gore (of "It's My Party"/"You Don't Own Me" fame).[55] Its cast also included the other three Firesign Theatre members, Richard Paul, and Firesign insider Michael C. Gwynne.

Roller Maidens returns to the idea of composing a story from multiple simultaneous television broadcasts, varying the conceit by allowing the television characters to watch each other's channels and travel between them to meet. The story mainly involves three programs:

1. *The Television Mission*, a congenial religious program that features the music of Red Greenback and the Blue Boys, a country band fronted by Austin, whose songs refer to the album's plot and in some cases extend it

2. an episode of a private detective procedural, set in the recent past, called *Carhook!*

3. alluding to the culture of reruns, a classic sitcom called *The Regular and Ethyl Show*, which draws openly on the 1950s shows *I Love Lucy* and *The Adventures of Ozzie and Harriet*

Involving, as it does, the oil crisis, glowing lizards, strange balls of light, and a midair wrestling match between Jesus and Richard Nixon, the complex plot of *Roller Maidens* can only be cursorily represented here. Its principal storyline is told through characters from vintage sitcoms, Juicy and Tricky Ritardo and Regular and Ethyl Boinklin, who find themselves in a supernatural mystery cued by the realization that *"we all seem to be dreaming the same dream"* and the rumor that "all the [television] announcers had quit, and moved, and there's nobody to introduce us anymore." Tricky and Regular additionally discover that their wives have absconded to a mysterious event at a country club, something they suspect has to do with Juicy and Ethyl's devotion to their women's club, the Mystic Roller Maidens from Outer Space. This leads them to contact the private dick Dick Private, a few channels away on *Carhook!*

What follows is a numinous variation on the familiar sitcom plot pitting the men against the women, as we discover that—wait for it—the husbands and wives have actually dreamed different dreams. Whereas the husbands' dreams have all been the expected erotic dreams about hot women, Juicy and Ethyl's dreams have been dreams of empowerment. Juicy describes the "happiest feeling, just like that dream was going to happen all over again," as Ethyl muses wanly, "Makes me kinda sad, you know, we Roller Maidens will soon control the world." Disguised in drag, Tricky and Regular follow their wives to the country club, where they observe the strongly hyped Nixon-Jesus wrestling match, which begins violently and then, once they are joined by the Roller Maidens, becomes erotic. They finally ascend into space, an event narrated by the stoic Dick Private, who has been following behind.

Roller Maidens' most intriguing implication is that the discovery that all the broadcasts are connected, and can be traversed as a single world, is related to the discovery that the male and female characters are respectively "dreaming the same dream." And here it is worth thinking harder than Austin may have done at the time about that idea's implications. From our perspective, the idea recalls Williams en route to Stanford, crashing in his Miami hotel with the TV on and receiving the fitful impression of "incidents as happening in the wrong film, and some characters in the commercials as involved in the film episodes." This famously led to the theory of "flow," in which a long sequence of programming (including ads) was understood to have a single organizational logic but where "the real organisation is something other than the declared organisation."[56] The shared dream in this sense might correspond to the logic of programming, extended across all the channels as a property of "television."

At the same time, however, it is significant that on *Roller Maidens*, the dream does not primarily belong to the viewer of television (sacked-out Williams or channel-surfing George Tirebiter) but to the characters within the drama, for whom it is either (for the men) the predictable raunchy dreams or (for the women) dreams of revolutionary empowerment, which the narrative suggests were false and manipulative. In this case, *Roller Maidens'* dreams are much closer to what Theodor Adorno called television's "dreamless dream"—a fantasy image that forecloses the possibility of empowering dreams, since "the stories teach their readers that one has to give up romantic ideas [i.e., to dream of a better life], that one has to adjust oneself at any price, and that nothing more can be expected of any individual."[57] This process of adjustment to degrading everyday conditions operates through a system of what Adorno famously called *pseudorealism*, the lesson that one must "be realistic," the recurrent moral of the classic television genres. Though the Roller Maidens' fantastical ascent into space is an escape from domesticity, it is rather different from the dream they began with, which was to control the world.

Adorno conducted his assessment of television in the Los Angeles of the 1950s (where he briefly returned in 1952–53, after having returned to West Germany in 1949), beginning with a rigorous content analysis of period sitcoms like *Our Miss Brooks* (CBS, 1952–56).[58] His "How to Look at Television" provides a particularly apt frame of reference for *Roller Maidens* not only for its mid-1950s source material (by that time accessible as reruns on 1970s LA TV) but also as an ambiguous document of the contemporary nostalgia boom. This was exemplified preeminently by the throwback 1950s-style sitcom *Happy Days* (ABC, 1974–84) and its many spin-offs. Ironically, *Happy Days* and its progeny created a televisual world its characters traversed in frequent guest appearances on their respective shows, a mundane and ironic normalization of *Roller Maidens'* imaginative cross-program travels. Although *Roller Maidens* undoubtedly lacks the affectionate nostalgia of the ABC sitcoms, it also conspicuously avoids another contemporary television source: the progressive social realist sitcoms of Norman

Lear, beginning with CBS's *All in the Family* in 1971. Things would be different when Austin thought seriously about television again in 1975.

AMBIENT TELEVISION, NIGHTMARES

The Firesign Theatre's apocalyptic final Columbia album, and most compelling statement about television, is 1975's *In the Next World, You're on Your Own*. In the history of the Firesign Theatre, we are slightly skipping ahead. In late 1973, more or less contemporaneously with the *Roller Maidens* recording, the Firesign Theatre agreed to make a new album, an expansion of their old Sherlock Holmes parody retitled *The Tale of the Giant Rat of Sumatra*, which they recorded at the RCA Studio in Hollywood. They then undertook the tour that was capped by the sold-out May 1974 Carnegie Hall show. In August, they went to the Burbank Studios (on the old Warner Brothers lot) to record *Everything You Know Is Wrong*, an album that would be accompanied by a quadraphonic mix (the first since *Bozos*) and a surprisingly effective dramatic film version, lip-synched to the album. For all their differences, *Giant Rat* and *Everything You Know Is Wrong* both foregrounded a defined central character, something that could not quite even be said of *Bozos*. The latter was a topical meditation on conspiracy theories and esoterica—with sources in Erich von Däniken, Uri Geller, and Carlos Castaneda—and has maintained an enduring reputation among listeners critical of, and friendly to, those cultures. It is their most extensively sampled album, and it is great. But although it contains numerous representations of television broadcasts—the nightly news, a local talk show called *The Golden Hind*, live coverage of Reebus Caneebus (Evel Knievel) and his "leap to the center of the earth"—it can't be said to be about television in the way that *The Martian Space Party*, *TV or Not TV*, or *Roller Maidens* are.

In the Next World, You're on Your Own, on the other hand—recorded in March 1975 on twenty-four tracks at the Burbank Studios—is much more impressively about television, drawing on myriad genres that include the Norman Lear sitcoms, soap operas, telethons, game shows, the Academy Awards, and the groundbreaking first reality program, *An American Family* (PBS, 1973). It also depicts the reception of television broadcasts in social space—what has been dubbed "ambient television." With Proctor and Bergman committed to touring their more accessible comedy, Ossman and Austin wrote the script (the first time all four had not collaborated in the writing). An enormous amount of notes, drafts, and research materials for *Everything You Know Is Wrong* and for *In the Next World, You're on Your Own* describe a world much larger than the one in which the albums explicitly take place but is still audible in the margins. For this, and for its ensemble performances and astoundingly detailed production, *Next World* is identifiably a Firesign Theatre record.

Everything You Know Is Wrong and *In the Next World, You're on Your Own* both take place in the fictional desert towns of Heater and Hellmouth, California, on the Arizona border. Consciously or not, the removal to the desert may also have been a way for the group to turn away from the racialized heteroglossia of their LA-based albums. At the same time, however, it afforded an opportunity to return to the scene of the very first Firesign recording and its bleak satire of Native American dispossession. Among Ossman's notecards detailing the record's prospective locations is one for "Strawberry Camp 13: 6 mi east of Fort Stinkin' Desert on S.D. Road. Now site of abandoned Historical Marker, once the wartime home of 100s of Los Angeles area gardeners and small businessmen during World War II."[59] This directly cited the Stinkin' Desert of "Temporarily Humboldt County," where the Indians had been first exiled and later removed for the creation of a nuclear test site, and this context would be important for *Next World*, too. Ossman added to that history its past as a Japanese-American internment camp ("gardeners and small businessmen"), further framing the desert not as a place of consoling white exurban escape but as the historical site of racial violence.

The working title of the album had in fact been "Sex and Violence." In an interview with the *Berkeley Barb* in November, Austin and Ossman stressed, "This album talks about especially television dreams. Those are the drives that are driving most people . . . so we watched everything that was available to watch on television that had to do with America's love affair with violence."[60] Readers of the *Barb*, though, would also have noticed that the interview—a two-page spread at the center of the paper— was bisected by a sixteen-page pull-out "adults only" ad section, filled with dozens of commercial ads and hundreds of personal ads for sex. By the mid 1970s, the underground press was almost entirely underwritten by the emergent sex industry. These are not obvious topics for a comedy album, as the following year's *Taxi Driver* would confirm.

Next World is an album with two distinct sides, the first side written by Austin and the second by Ossman. They differ formally, but they share a single space and time and intersect at two critical points. Indicating how this was a part of the album's design, the manuscript materials include annotated drawings of street maps, with X-marked intersections indicating both the crossing of the two narratives and the traffic intersection of "Police and Eat Streets," which will be the site of a fatal car collision (fig. 28). (Pretty funny so far, right?) These crossings are made explicit in William Stout's superb cover art which depicts a cop's black-and-white screamer on a collision course with a 1940s convertible, cars driven by characters from each of the record's two sides (fig. 29). Described on both sides of the album, the car crash is the first of the album's two narrative crossings. The other crossing would be explicitly televisual.

Side 1

Austin's side of the album is, in a sense, told entirely through television. It cites the sociorealist *All in the Family* by telling the story of a lower-middle-class white family that loosely resembles *All in the Family*'s Bunker family: there is a working father, a stay-at-home mother, an adult son and daughter. Austin multiplies the television references by dramatizing each of the characters' stories, not in the language of realism but through television genres demographically associated with each of the characters.

- The father, Random Koolzip, is the main character on a cop show, *Police Street*: "It's the worst street in town, it's *so bad*."
- The mother, Peggy, is the lead on the soap opera *Over the Edge*, "the story of Peggy and what a woman feels when she must find out for herself the things her husband does not know."
- Kim Kool, their daughter, is cohosting the celebrity telethon *Circle of Excitement*, where "we're climbing out of Hell on your phone calls."
- Skipper, the son, is a contestant on the game show *Give It Back*, which wrests away his family's possessions in a version of reparations for American Indians: "Skipper will be accumulating prizes for this week's Abandoned Americans—the Dog Indians."

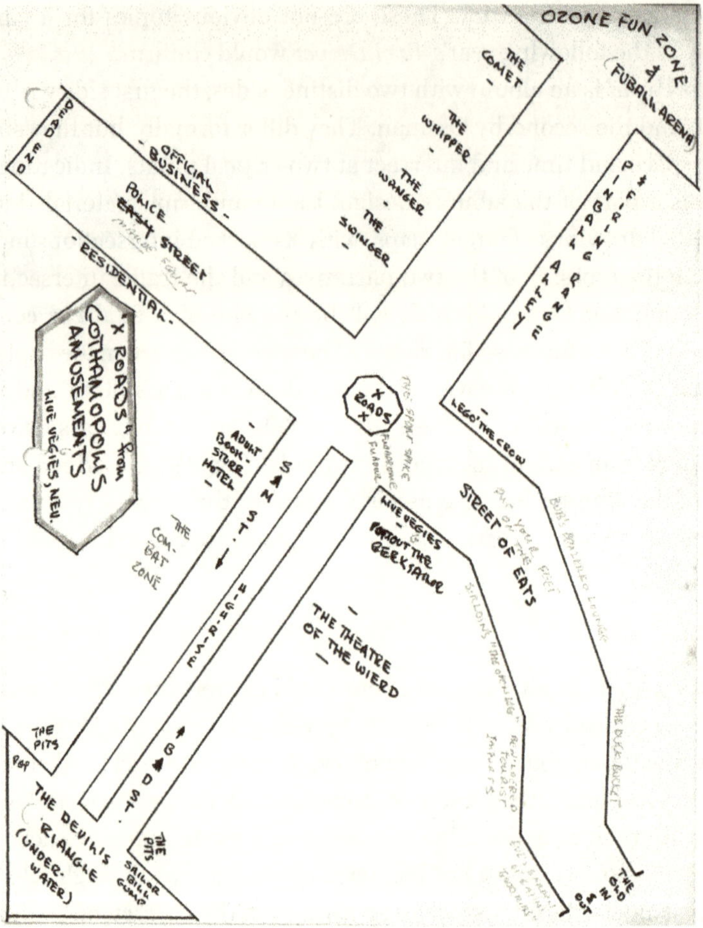

Figure 28. David Ossman's map of Heater, California, showing the X intersection at Police and Eat Streets. Draft material from *In the Next World, You're on Your Own.* Firesign Theatre Collection, National Audio-Visual Conservation Center, Library of Congress.

With the exception of a short police radio conversation on *Police Street* between Random and Peggy, none of the characters speak to each other, and they are never physically in each others' presence (the antithesis of what would normally drive the plot of a family drama or sitcom).

The extent to which these stories are meant to take place "on television" is pointedly ambiguous; there is no metafiction of a television and implied

Figure 29. William Stout's cover art for *In the Next World, You're on Your Own* by the Firesign Theatre. Courtesy of William Stout and Sony Music Archives.

viewer, as there is on Firesign's other TV albums. The writing, characterization, and sound design nevertheless all emphasize the generic specificity of each character's world, cues clearly given by the discrete vocal performances, the voice-over announcers, and the overwhelming use of library music conveying mood. When Peggy talks to Random in the side's first vignette, she appears as an adoring wife who sees her husband as the "crumpled kind of hero" described by *Police Street*'s announcer. But in *Over the Edge*, we learn that Peggy has fallen in love with her friend Eliot, who can be heard muttering that Random is "just driving around, drunk probably" before Peggy screams that "Random's never coming home!"—predictions that are both realized in later sequences of *Police Street*.

The overall concept can be elucidated by referring once again to Raymond Williams, in this case his mid-1970s lecture "Drama in a Dramatised Society." Here, Williams begins by broadly observing that because of television, "for the first time [in human history] a majority of the population has regular and constant access to drama," pointing out that most people now spend far more time "watch[ing] simulated action" than they do eating (a distinction Firesign would have appreciated).[61] In comparison, he points out that "in earlier periods drama was important at a festival"—such as, say, during the medieval period of carnival—an experience that would have been temporally delineated by the calendar. Because "drama, in quite new ways, is [now] built into the rhythms of everyday life," "what we often have now is a new convention of deliberate overlap" to such an extent that we are left "continually uncertain whether we are spectators or participants." The ambiguous status of these characters on *Next World* is an expression of this same uncertainty: though their vignettes are highly artificial, Random, Peggy, Skipper, and Kim are not obviously characters on TV.

Firesign's professional actor, Phil Proctor, had had a one-off role in a 1971 episode of *All in the Family* (November 27, "The Insurance Is Canceled"). And as it happens, there was a separate personal connection informing still another of the television references on *Next World*'s first side. The Ossmans knew their fellow Santa Barbarans the Loud family, who were the subject of the pathbreaking docuseries *An American Family*. *An American Family* is specifically, if obliquely, referenced within the text of *Next World* in the form of a crisis concerning the son, Skipper, in the *Over the Edge* segment. Part of the humor of *Over the Edge* is that Eliot and Peggy's intimate conversation is repeatedly interrupted by an escalating number of nosy house-callers. Wafting through this excruciating scene is the rumor that Skipper has been found by his father together with two other boys, "poor little Bozo and Bill Bag" ("doesn't he wear . . . *eye make-up?*"). As a way of introducing this rumor, the neighbor Sue pleads, "Is it true? Is it true that Skipper can't be Governor now? Not ever?" Though here dramatized as gossip that Peggy is obliged to confront and discuss—it notably is not referenced elsewhere—it is clearly a reference to the most famous event in *An American Family*: Lance Loud's coming out as gay to his mother Pat. *Next World* noticeably sidesteps what television scholar Amy Villarejo calls the program's "stunning" "privileg[ing of] Lance's queer epistemology," a landmark event in US television and culture.[62] Skipper, when he appears in the game show *Give It Back*, will lack any of the flamboyant queerness that distinguishes Lance in the program. In place of that recognition, we hear Peggy's exasperated defense—"Skipper's better off the way he is!"—which is in its delimited way admirable. But it fails to endorse the project of this early reality program as a whole, about which the anthropologist Margaret Mead wrote (in the pages of *TV Guide*), "[*An American Family*] may be as important for our time as were the invention of drama and the novel for earlier generations: a new way to understand themselves."[63] Firesign seems more inclined to understand reality television as the latest of the conventional genres, though it is worth noting their acknowledgment of *An American Family*'s importance.

Side 2

Another notable feature of the pessimistic *In the Next World, You're on Your Own* is its disappointed skepticism about 1960s political causes, something that the album tied directly to the phenomenon of the media. This entailed a significant amount of self-assessment and critique. It can be heard, for instance, in the radio advertisement for "Billy Jack dogfood—you know, the dogfood Billy Jack likes," a withering dig at the eponymous 1971 Warner Brothers movie about a mixed-race Navajo martial arts master and Vietnam veteran who defends a countercultural Freedom School in Arizona. (As Roger Ebert would write, "[*Billy Jack*] has as many causes in it as a year's run of the *New Republic*."[64]) But it also named political concerns about which Firesign had cared deeply.

Their disappointment is audible in the storyline that most noticeably traverses the album's two sides, beginning on side 1 with Skipper's appearance on the game show *Give It Back*. *Give It Back* produces as entertainment the process of Native American reparations, and its gaudy display of gleeful bad faith is precisely Firesign's point. Presided over by deliriously amped up hosts, *Give It Back* is clearly modeled on contemporary shows like *Let's Make a Deal*, which *Rolling Stone* would profile later that year as "humiliation games."[65] As its willing contestant, Skipper is made to wear a war bonnet and "fantastic, bullet-proof, Indian Ghost Dance Shirt" before being sent into the program's Gitchee-Goomee Money Tunnel, which is "blowing up—a hurricane—of *silver dollars!*" that Skipper is meant to collect while covering his face with his shirt. The first stages of the game involve the distribution of his parents' cars to the grateful Dog Indians. When he protests that they'll need the car, the announcer shouts smugly, "Relax and *be an Indian!* Forget about the restless world of Anglo-Saxon discipline, Skipper, because it's *round two!*"

The grand prize, however, has to do with publicity. Skipper is to be awarded "a dream package to the capital of dreams, Hollywood, California" to attend "the fabulous . . . *Academy Award Celebration!*" where he and his sister Kim will "represent your tribe at the very citadel of power" (though they are not Indians), "mingle with the attractive crowd," "introduce yourself to important people who may be able to help *your tribe* back home," and "step backstage to snort cocaine with a well-known entertainment personality." Skipper is then enlisted to make a speech:

KEN: It's your chance to lift your people out of bondage as you brandish a real double-o-gauge shotgun by Bangge of Brentwurst!

SKIPPER: Hey!

DON: Skipper, you'll address the glimmering crowd of stars *like an equal!*

SKIPPER: That's a gun!

DON: And you'll hold the Nation spellbound with these stirring words:

KEN: "Eat flaming death, fascist media pigs."

GENE: Yes, siree! You'll hold *all* of Hollywood for *ransom*, Skipper! . . .

DON: Yes, Skipper—it's your gallant bid to give America back to the Dogs!

We see how this turns out on the album's second side.

Written by Ossman, the B side of the record follows an investigator protagonist named Sugar Burns, who has come to the high desert to investigate an Indian labor uprising, according to the script drafts. After he is hazed by a local cop who advises him that "Marlon Brando or some crazy Indian is using the interstate as a shooting gallery," the narrative is given over to Burns's car radio, on which we first hear a minor-league baseball game (between the Locusts and the hometown Crows) before it tunes in to the Academy Award celebration. Here, we hear magnificently detailed clips of the five films nominated for "overachievement in a starring vehicle": Sir Leonard Stove in *Ten Dead Bats*, the late Victor Kabloona in *S.Q.U.A.T.* (which turns out to be a glossy film version of *Police Street*), Charlton Heston in *The Day Universal Burned Down*, Ace Berzerker in *Crime Clock Terror*, and Frank Clearwater as Ghostdancer in Oklahoma Mutoscope's *I Gave It Back*. Confirming that the fix was in, the award goes to Frank Clearwater, and the award is accepted by Skip Zipper, costar of television's *Police Street* and his sister, Kim Kool, recently featured in *Daddy's Biggest Girl*. As foretold, they hold the audience hostage: "I want the President of the United States, a plane full of cash, and all those broken treaties!"

Amusing as this may be as fiction, its politics are profoundly equivocal, as Skipper's muffing of the speech indicates ("Eat fascist death, flaming media pigs!"). This is true because of the way the scenario conjoins allegorically two recent events: Sacheen Littlefeather's appearance at the March 1973 Academy Awards, where she accepted Marlon Brando's Best Actor Oscar during the American Indian Movement's (AIM) occupation of Wounded Knee at the Pine Ridge Reservation in South Dakota;[66] and the February 1974 abduction of Patricia Hearst, granddaughter of media mogul William Randolph Hearst, by the militant Symbionese Liberation Army (SLA), with whom Hearst would participate in an armed bank robbery in April. Returning to their own most ardent political cause—redressing Indigenous dispossession—the Firesign Theatre question the degree to which media, and television in particular, had exploited and neutralized it. When Kim Kool addresses the crowd, it is in the terms that recall the triumphant conclusion of *How Can You Be in Two Places at Once*: "Brothers and sisters! There is only one way to win this war! Surrender!" But here it is the booth announcer who has the last word: "Well, it's a moment of high ratings and suspense among the glamorous and talented crowd gathered . . . for tonight's annual celebration. We'll return—after this."

Ossman's working papers for *Next World* contain pages and pages of notes about violence (a list of subheadings includes "sports," "crime," "revenge/self-protection," "mass violence," "religion," "violent sex," "emotions," "effects," "comic violence," "fantasy," "tv news," "entertainment," "public policy," and many others). At the top of one page are two epigraphs: "Violence is as American as apple pie" (H. Rapp Brown, 1967) and "Collective violence is normal" (Charles Tilly, 1969). The University of Michigan sociologist Tilly is further quoted from his contribution to the 1969 report *Violence in America*: "the stimulus toward collective violence comes largely from the anxieties men experience when social institutions fall apart."[67]

These ideas were refined and adapted throughout the process of composition, as is evident in the several connotations that would be attached to the concept of the "Next World." That phrase first appears in Ossman's draft materials as the name of an "X-Rated Family Amusement Park" in nearby Hooker, California, an idea William Stout's cover illustration adapts as a billboard advertising the "Next World Amusement Park and Burial Ground." But the Next World concept is given a different life in the apocalyptic conclusion of the album, where it is also associated with the album's only explicit representation of a television set.

Turning off his car radio, Sugar Burns drives into town where he notices a liquor store—Ma's Brontosaurus Liquors, which is housed in "a giant cement lizard all lit up inside." There is a TV behind the counter—an example of what media scholar Anna McCarthy has dubbed "ambient television"—on which we can hear the Crows-Locusts baseball game that had earlier been heard as a broadcast on Burns's car radio.[68] If we listen carefully, it appears that the baseball players are actually birds and insects ("[Mickey] Mantis backpedaling ... of course, he's got that 360 degree vision ..."), and we may be leaving the world of realism in other ways, too (since Ma tells us that the championship game is "draggin' on for an extra week"). Burns approaches the counter to buy condoms and a bottle of whiskey but before he's able to pay, a shotgun blast rips into the store and shatters the TV. Burns and Ma hide behind the counter. Burns whispers, "It's not just happening here, it's happening in Hollywood, I heard it on the radio!"

MA: This is a strange place. Just as strange as Hollywood, any day. I expect—you've heard—*the Voices.*

SUGAR: No. What Voices?

MA: Oh, most people hear 'em as soon as they come into town. You gotta keep your TV on real loud all the time to drown 'em out.

VOICES: Aaargh . . . Sugar sugar sugar . . .

SUGAR: Oh! Is that one?

MA: They must come up from The Next World—that's all we know!

VOICES: Sugar sugar sugar.

SUGAR: How does it know my name?

MA: What? Oh they call everybody "Sugar." Be careful! Don't let 'em read your mind. Sing! Sing something! Anything! . . . Come on, sing, or they'll getcha!

The grotesquely commercial Next World of the early drafts appears here as the apocalyptic manifestation of transfiguring spirits. Ma and Burns try to drown out the Voices by singing and by turning up the radio broadcast of the game, where we hear that a million locust eggs have hatched from under home plate and a butterfly has flown out of the back of the back of another insect's uniform "and it's all-out warfare down on the field." Burns's escape from this frightening, swirling sound world is simply to shout STOP. And the scene inexplicably relocates to a burbling stream where Burns—concluding that it's "Too hot in Heater—too close to Hellmouth—Too much life and death over in Hollywood"—casts a fly rod and suggests "there's more than one way out."

This magically consoling, but ultimately unconvincing, resolution to the record brings to mind nothing so much as the philosopher Stanley Cavell's challenging 1982 essay "The Fact of Television," which takes as its impetus something that could not be said of the Firesign Theatre—namely, "the absence of critical or intellectual attention to television."[69] Cavell's argument is that the serial format of fictional programming, no less than the more obviously time-bound broadcasts of news or talk programs, discloses "the material basis of television as *a current of simultaneous event reception*" (85), which is to say that the experience of its reception is that of "*monitoring*," a disenchanted rephrasing of the more familiar understanding of the everyday omnipresence of television as "providing 'company'" (86). This is offered as a property of the television itself: "if the event is something the television screen likes to monitor, so, it appears is the opposite, the *uneventful*, the repeated, the repetitive, the utterly familiar"—which no doubt is why the people of Heater take consolation in their TVs as a way of drowning out the intrusive supernatural Voices whose origins are obscure but undoubtedly threatening: "To find comfort or company in the endlessly uneventful has its purest realization, and emblem, in the literal use of television sets as monitors against the suspicious, for example, against unwanted entry" (89).

Acknowledging the centrality of television to coverage of the civil rights movement and the Vietnam War, Cavell asserts that the originary moment of television's "conquering" of all the other media—and of everyday life— had nevertheless already occurred at the moment to which the Firesign Theatre would also repeatedly return: after World War II, "after the discovery of concentration camps and of the atomic bomb; of, I take it, the discovery of the literal possibility that human life will destroy itself; that is to say, that it is *willing* to destroy itself" (95): "Not to postpone saying it any longer, my hypothesis is that the [critical] fear of television—the fear large or pervasive enough to account for the fear of television—is the fear that what it monitors is the growing uninhabitability of the world, the irreversible pollution of the earth, a fear displaced from the world onto its monitor (as we convert the fear of what we see, and wish to see, into a fear of being seen)" (95).

From the perspective of our own hearing of *Next World*, it is impossible not to recognize the aptness of the album's desert location—peripheral from the cultural centrality of Los Angeles, home to the racialized violence of Indian genocide, and central for our purposes in the new era of climate change. A month before I wrote these lines, the city of Phoenix endured nineteen straight days of 110-degree temperatures.

Just as Streeter foresaw the way the Blue Sky discourse promoting cable television would be reframed by the techno-utopian discourses promoting the web, rereading Cavell's 1982 essay also calls us to attend to the twenty-first century phenomenon of internet culture, where mobile screens provide both a daily monitor of reassurance and an actual monitor of Earth's "growing uninhabitability." It is in this intellectual context that the Firesign Theatre's television meditations, though their earlier records provide a dizzyingly inventive media archaeology that provides genuine historical and counterhistorical knowledge, are the location where they find their fullest realization as jocose Cassandras. No wonder their contract wasn't renewed.

CODA / Run-Out Groove

NOT TV AND NOT ROCK EITHER

To put it another way, the moment the Firesign Theatre lost their Columbia Records contract was also the moment television—the medium about which they had thought so variously and to which they had the least access—became the most important platform for comedy in the United States. NBC launched *Saturday Night Live* in October 1975 (featuring veterans of the *National Lampoon's* aspirational Woodstock parody *Lemmings*), and it soon became a crucial venue for rock musicians and for stand-up comics. Nineteen seventy-five was also the year of Monty Python's American apotheosis, as *Monty Python and the Holy Grail*'s huge box-office success capitalized on the previous year's US syndication of their BBC TV show. And it was the year the cable network HBO premiered, the first of its influential stand-up comedy specials broadcasting at year's end, inaugurating a form that still thrives today on Netflix and elsewhere.[1]

Though innovative multitracked comedy albums continued to be made in Firesign's absence—Lily Tomlin's *Modern Scream* and Albert Brooks's *A Star Is Bought* were both released in 1975—most of these artists had access to television, too.[2] Brooks's short films were regular features on the

201

first season of *Saturday Night Live,* and Tomlin had already won an Emmy, after having broken through on *Rowan & Martin's Laugh-In* (1968–73). She would host *SNL* in 1976. Between the two of them, they only made one more album, Tomlin's more conventional *On Stage* (1977). Concert recordings like *On Stage,* meanwhile, came increasingly to dominate the field, led by the mid-1970s albums of Richard Pryor, George Carlin, and the apolitical, arena-friendly Steve Martin.[3] All of these stand-up artists would also appear on *SNL,* Martin hosting the show seven times before 1980.

More than any other nonmusical act, the Firesign Theatre had been accepted as a part of album culture and had even in a sense been understood as making a kind of music. On the cusp of recording *Led Zeppelin III,* Robert Plant had said that "young people can get into all sorts of music; my idea of 'heavy' . . . is more Incredible String Band and Firesign Theatre and things like that."[4] So it may not be surprising that the Firesign Theatre were never offered a major break on network television; Cheech and Chong were never offered one either.[5] But whereas Cheech and Chong savvily maintained a presence on FM radio with rock and soul tracks like "Basketball Jones" (1973), "Earache My Eye," and "Black Lassie" (1974), the Firesign Theatre soon found itself both excluded from television and marginalized from the music culture in which it had once been preeminent.

The latter was epitomized most painfully by the group's evolving fortunes in *Creem,* the publication that had supported Firesign most ardently. Jaan Uhelszki (famed author of "I Dreamed I Was Onstage with KISS in My Maidenform Bra") has recalled how legendary writer-editor Lester Bangs would affectionately describe the *Creem* writers as "just 'bozos on this bus.'"[6] And Bangs indeed began his long 1975 profile of the Firesign Theatre by recalling how "a little bit of boo" and side 1 of *How Can You Be in Two Places at Once* had once "made [his] psychic synapses twang like a dobro."[7] By now, however, Bangs admitted that he "no longer had the patience, attention span, consciousness-level, or whatever was required to 'get into' Firesign Theatre albums in the classically accepted fashion." ("Putting it bluntly, I was now a drunk.") And he then commenced a notorious dethroning in the spirit of his legendary Lou Reed portfolio that mean-spiritedly, if accurately, emphasized how annoying Firesign could be in interviews (*vide* Phil Austin's "equine smugness") while acknowledging in passing the Firesign Theatre's estimable vocal performances and sui generis contributions to the representation of media.

Bangs's agon laid the ground for a neurotically equivocal review of *In the Next World, You're on Your Own*, penned by the same critic who had written *Creem*'s first vatical Firesign reviews in 1969. Though Richard C. Walls ranked *Next World* among the Firesign Theatre's best work, he also felt obliged to apologize for (among other things) the group's "predominant and frequently irritating" use of puns and conceded with (I guess) remorse that drugs might be desirable or even necessary for comprehending the long dense works. Doubtful of winning converts, Walls concluded by insisting that any new listener would first need to hear all the preceding albums in order; otherwise, "it's like . . . picking up John Coltrane's record *Om*, listening to it and deciding that he's a ruse without checking out the fifteen years of recorded development that led up to it. It really is."[8]

The most brutal instance of *Creem*'s volte-face came in 1981, when Dave Marsh, whose ecstatically paranoid appraisal of *I Think We're All Bozos on This Bus* had headed the magazine's review section in November 1971, failed to include Firesign among the twenty-one "Famous Comedy Rock Groups and Performers" in his *Book of Rock Lists*. Making the cut were bands and comedians such as Cheech and Chong (4), Monty Python (10), the Bonzo Dog Doo-Dah Band (2), Blowfly (1), the Tubes (20), the Fugs (8), the Plasmatics (15), and (yikes) Sha Na Na (19).[9]

Bangs's disavowal, Walls's prevarication, and Marsh's oversight were each evidence of significant shifts in the production and reception of records and, more broadly, in practices and sensibilities related to technology, drugs, politics, and the culture industries. There had been, for instance, a salutary diversification and semidemocratization of the airwaves, as new radio formats targeting discrete demographics began to proliferate.[10] The Firesign Theatre, whose work had made the most sense on adventurously ecumenical playlists, did not benefit from this transformation. Their consequent exile from television, radio, and major-label records meant the group had lost access both to the tools and the audiences of mainstream media. In 1976, Columbia saw out the group's contract with a valedictory double-LP anthology, and that was it.

The Firesign Theatre nevertheless continued to work together, albeit with fallow periods, for another twenty-five years (see appendix A). During these years, they produced ten more studio albums (some of them excellent), forays into home video, stage performances, and in 2001 revived their live radio program on the XM Satellite network. But though these often involved their characteristic curiosity about technology, for instance in their mid-1980s interest in video games, Firesign Theatre's post-Columbia work was not media archaeological in the way that had so distinguished their earlier albums.[11]

Those records had been made possible through the patronage of a major recording studio situated among the other intersecting media industries, all of them in periods of profound transition. They were also the product of a sensibility and technique of collective improvisation that their KPFK colleague Bill Malloch had stressed was "a matter of ongoing temperament"; this proved much harder to maintain once the group were no longer making a living from their records (also, they were getting older). Interdependent as they were, the fact that these institutional and temperamental conditions—which Walter J. Ong might have called "technological" and "oral"—were transitory is one way of understanding why the Firesign Theatre would have no direct inheritors in the creative form they invented.

CAMPOON '76

They would have many prodigious indirect inheritors, however. As the Firesign Theatre's Columbia albums rapidly entered the domain of minority culture, they became the inheritance and archival source of many forms of underground culture, which included autonomous work of fan communities, the critical audio work of groups like Negativland, and the esoteric expertise of several generations of hip-hop DJs. To briefly trace these genealogies, we should begin with the remarkable project engaged across great distances by Firesign fans at the moment it became clear that the group did not have a major label future. Produced with the cooperation and encouragement of Ossman, and with some participation from the rest of the group, it was a monthslong multimedia project that hearkened back to the participatory commedia dell'arte of the Renaissance Pleasure Faire.

The first signs of an independent fan community could be seen as early as 1971, when informal classes devoted to the Firesign Theatre began springing up at "free universities" in college towns like East Lansing, College Park, and Oshkosh, followed by more official offerings at places like UCLA, Indiana University, The New School, and San Diego State.[12] These in turn anticipated stage productions of each Firesign album (abetted both by the publication of the scripts and by fans having memorized the albums) that appeared at theatres in Amherst, Iowa City, Tucson, Santa Fe, Ashland, and elsewhere. The Firesign Theatre had by then begun to involve the fans in their work, hailing them as the participant-delegates of *The Martian Space Party*'s presidential convention (as discussed in chapter 5). A half-page précis on George Papoon's campaign for president appeared in *Creem*, followed by interviews with Proctor and Bergman on behalf of their imaginary candidate, which Columbia distributed to radio stations as a white-label record, timed to the 1972 election.[13] As fans began to participate in Firesign's fictions, the Firesign Theatre experimented with extending those fictions into the world.

By then, the idea of a self-organizing fan group had also emerged in the form of the first independent fanzine devoted to the Firesign Theatre, *Firemail*. Founded in Lincoln, Nebraska, by Tom Gedwillo, *Firemail*'s list of subscribers had been composed from responses to a short classified ad in *Rolling Stone*. Expanded and rebranded as *Chromium Switch* the following May, the fanzine allowed fans to connect across distances (through the mail) and communicate semi-independently, the project gaining steam just as announcements of Firesign's breakup were appearing in *Rolling Stone* and *Rock* magazine. Picking up on the presidential convention plotline of the *Space Party* and *Not Insane*—and further integrating the Papoon candidacy into society—*Firemail* advertised fan-produced paraphernalia (Papoon for President bumper stickers and pins), which no doubt is how the buttons reached the lapels of John Lennon in 1973 and (according to Edgar Bullington) Emmylou Harris in 1974.[14] All of this began to formalize the performance of fandom that had hitherto been distributed covertly in places like phone-book entries for Nick Danger and recipes published under the name of Mrs. George Tirebiter (see preface).

The Papoon for President concept was realized more fully in the next election cycle, however, directly engaging hundreds of Firesign Theatre fans in an example of what the media scholar Henry Jenkins has called "how texts became real." Jenkins's seminal 1992 study revalued the work of neglected everyday consumers, focusing on media texts like *Beauty and the Beast* and *Starsky and Hutch* that are perhaps the antithesis of the ambitious literariness of the Firesign albums. But his three dimensions of participatory television culture nevertheless all describe the practices of Firesign fans from the 1970s: "the ways fans draw texts close to the realm of their lived experience; the role played by rereading within fan culture; and the process by which program information gets inserted into ongoing social interactions."[15] In the 1976 Papoon campaign, these resulted in a very large multimedia coauthored mass art project that Ossman hailed as "grassroots surrealism," beginning in the autumn of 1975 and continuing through Papoon's celebratory inaugural ball in Santa Barbara in February 1977. The paper trail of this eighteen-month project now fills eight storage boxes at the Library of Congress.

The Firesign Theatre had floated a trial Papoon balloon in the summer of 1975, as they increasingly began devoting their three-page monthly spread in *Crawdaddy* to Papoon-themed materials (this was a side gig they'd had for a year). But by far the most extensive, varied, and interesting work was done not by the group but by the fans. The final item in the October *Chromium Switch* announced that "George C. Papoon's Natural Surrealist Party Campoon for President is shifting into low gear, with Cocoons spinning out propaganda all over the country." Though this proved to be the last issue of *Chromium Switch*, a new fanzine soon appeared that would serve as the information center for self-organizing Papoon Cocoons throughout the election year of 1976. Published from Topeka, Kansas, by Steve Cowell, *The Toiler* was a sixteen-page offset-printed newsletter that contained news items (its first issue announced "Columbia Drops Firesign"), mail from fans, nonsense articles by fans (and occasionally Firesign members), advertisements for Papoon memorabilia (balloons, iron-ons, membership cards), offers of Papoon-related services (e.g., airbrushed T-shirts bearing the name of your Cocoon), *legal* services from Firesign's lawyer Richard Shulenberg, and advice on staging Campoon events, including guidelines for press releases and communicating with the media.

Most important, *The Toiler* included a regularly updated list of the nearly one hundred Papoon Cocoons that formed across twenty-eight states (one on the Rosebud Indian Reservation in South Dakota), the

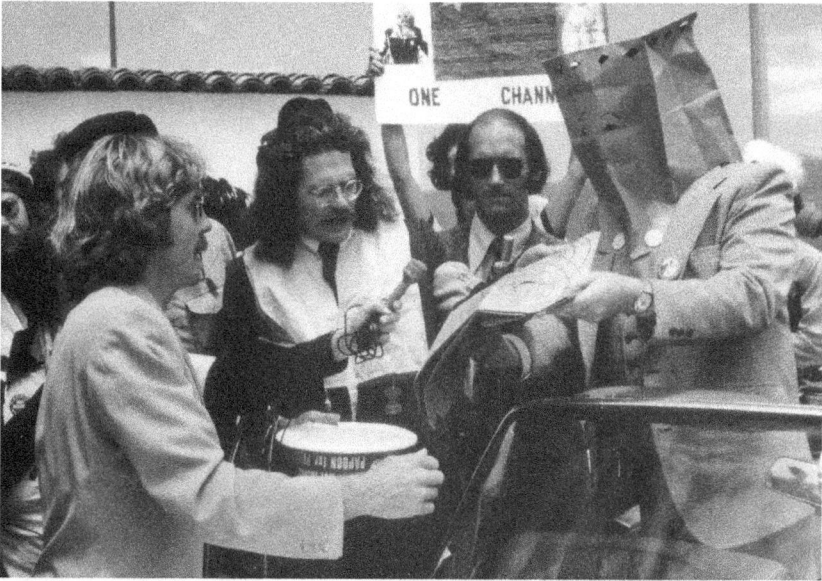

Figure 30. George Tirebiter (David Ossman), press, and supporters greet George Papoon at Santa Barbara airport, July 31, 1976. Photo by Patricia Pence. Firesign Theatre Collection, National Audio-Visual Conservation Center, Library of Congress.

District of Columbia, Ontario, and British Columbia. Some of the Cocoons had access to local FM stations and were able to append five minutes of Campoon information to a news broadcast. Many, or even most, produced their own fanzines and newsletters that were wholly, partly, or seemingly not at all related to the Firesign Theatre: *Is This 'Icrophone 'Orking*, from Seattle; *Doing the Rag*, from Provo, Utah; *Tips for Zips*, from Mineola, New York; the *Too Close to Be Newsletter* from Glendora, California; and David and Tiny Ossman's *Grape-Vine*, among many many others. The proliferation of Xeroxed and mimeographed materials provides a glimpse at an inflection point between fanzines' origins in the sci-fi culture of the 1930s to their mid-1990s apogee, just before the advent of the World Wide Web, chronicled and promoted by Mike Gunderloy's *Factsheet Five*.[16] (My own not-very-good first attempt at writing about the Firesign Theatre appeared in my and Dan Williams's zine *Verbivore* in 1994).

The Papoon for President campaign emblematized the political disenchantment of the post-Watergate era. (The much-diminished Yippies

Figure 31. Zippo Klein nominating George Papoon for president, National Surrealist Party convention, Santa Barbara. Photo by Patricia Pence. Firesign Theatre Collection, National Audio-Visual Conservation Center, Library of Congress.

launched their "Nobody for President" campaign that same election cycle.)[17] Whereas Papoon's 1972 run had been born under the despairing sign of Nixon's imminent landslide reelection and was produced largely by Firesign themselves, Campoon '76 was a product of the fans' otherwise disaffected creativity and testified to a desire for community and engagement. More than once, *The Toiler* issued reports of citizens running as National Surrealist Party candidates in municipal and student government elections.

In 1972, Papoon had been represented by a vaguely Nixonesque mask procured by Proctor from a Rexall in Studio City. In 1976, the mask was replaced by a simple brown paper bag cut with two holes for eyes. In neither case was Papoon directly associated with a Firesign Theatre member and he was never given one of their many voices. Perhaps accidentally, this maximized the possibilities for the participation of the fans, who could acquire all the necessary equipment for a Papoon appearance by making a

Figure 32. National Surrealist Party convention attendees, Santa Barbara. Photo by Patricia Pence. Firesign Theatre Collection, National Audio-Visual Conservation Center, Library of Congress.

trip to the grocery store. Such was notably evident at an unannounced Fourth of July Papoon meet-and-greet at a Southern California nude beach (dutifully documented in *The Toiler*). And it reached its fullest expression at nominating conventions in Santa Barbara and Lawrence, Kansas, both of which were multiday gatherings (figs. 30–33). Each was authenticated by Ossman's in-character appearance as VP nominee George Tirebiter, but they were also opportunities for all the fans to appear in character, as delegates, press and media figures (there was TV and radio coverage for both), Surrealist Party candidates, would-be assassins, and of course Papoon himself. More than the scripted audience participation that would later form around screenings of *The Rocky Horror Picture Show* (1975), the Papoon events not only made the Firesign texts "real"; they also inspired improvisatory participation in the absence of a directly experienced text.

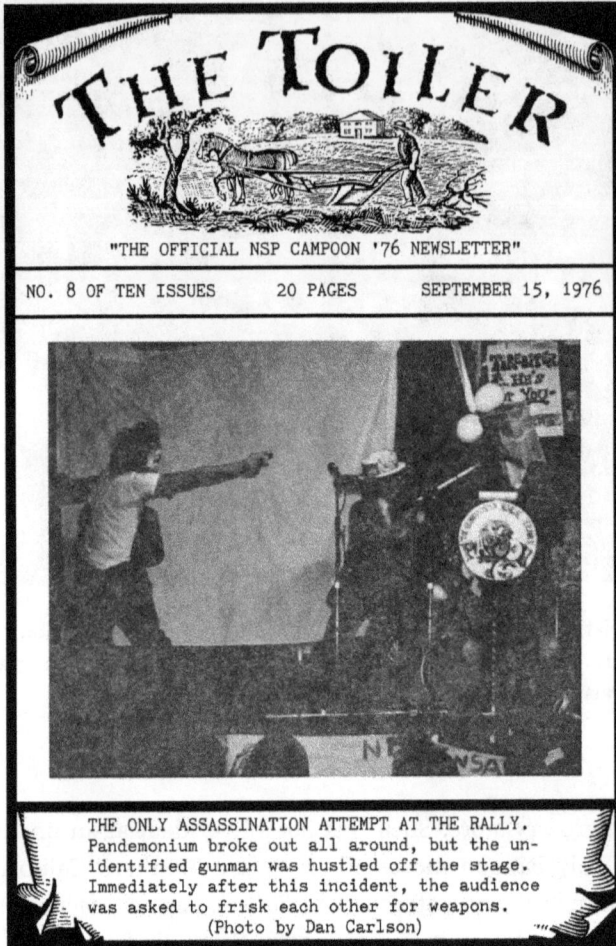

THE TOILER

"THE OFFICIAL NSP CAMPOON '76 NEWSLETTER"

NO. 8 OF TEN ISSUES   20 PAGES   SEPTEMBER 15, 1976

THE ONLY ASSASSINATION ATTEMPT AT THE RALLY.
Pandemonium broke out all around, but the un-
identified gunman was hustled off the stage.
Immediately after this incident, the audience
was asked to frisk each other for weapons.
(Photo by Dan Carlson)

Figure 33. The Toiler's cover story of "the only assassination attempt" at the National Surrealist Party convention, Lawrence, Kansas, September 1976. Photo by Dan Carlson. Firesign Theatre Collection, National Audio-Visual Conservation Center, Library of Congress.

THE CHURCH OF THE SUBGENIUS AND NEGATIVLAND

The '76 Campoon was generative in a way that forming a comedy group in Firesign's image would not have been. A notable institution that emerged directly out of the Papoon campaign, taking as its first mailing list *The Toiler*'s national directory of fan cocoons, was the Church of the SubGenius.[18] Representing the Dallas, Texas, Bulldada Time Control Labs Cocoon at the Lawrence, Kansas, Papoon convention was the twenty-three-year-old documentary filmmaker Douglass St. Clair Smith. Smith had been avidly involved in the Campoon from the beginning and contributed an informed "History of Surrealism" to the second issue of *The Toiler*. Three years earlier, Smith had written and directed one of the earliest and most impressive examples of Firesign Theatre fan art, a forty-five-minute black-and-white student film titled *Let's Visit the World of the Future*.

Still easily found on YouTube, *Let's Visit the World of the Future* is a savage and eerie period satire of US consumer culture, comprising short theatrical sequences intercut with stock footage of shopping malls, airports, hospitals, and factories. Smith's film both borrowed and extended the sensibility and iconography of *I Think We're all Bozos on This Bus* (which it cited in its credits) by involving clowns, holograms, "the future," and hyperreal environments as a part of its world. "I'll be the guide on this trip through reality," its voice-over begins, "and on this trip I can make you see anything I want you to see." It goes on to reveal a world "where humans no longer endure self-inflicted pains, where one is *made* to enjoy life. . . . Why, just looking at the future is enchanting: carouse along conveyor belts where awe-inspiring holograms are as much a part of the experience of every man as were the real trees, mountains, and sunsets of the dark past." But the film also adapts *Bozos* for its own purposes. Whereas Firesign's Bozos are the fundamentally passive hedonists hailed and emulated by Lester Bangs (the credulous but well-meaning attenders of the Future Fair), the clowns of *Let's Visit the World of the Future* are facetious, malevolent, and decidedly on the side of "society" (and possibly also inhuman).[19] Smith's clowns use giant syringes, for instance, to inject humans with a hallucinogenic placating drug called Binsky's Nuclear Beer. *Let's Visit the World of the Future* was an homage to *Bozos* but also a possible critique suggesting the album had not gone far enough. In refusing to be completely faithful to its source, it endowed the source with new life, and the fact that its citation was only available to the cognoscenti was also thoroughly in the spirit of the Firesign Theatre's mode of citation. With Ossman's blessing, *Let's Visit the World of the Future* was screened at the Surrealist Party convention in Lawrence along with other fan films and the Firesign Theatre's own movies.

Using *The Toiler's* fan directory to compose its first audience, Smith founded the Church of the SubGenius three years later with the inaugural *Pamphlet #1—The World Ends Tomorrow and You May Die!* The Church of the SubGenius (or the "Church of Bob") became known as an apocalyptic consumerist parody religion of "slack" that satirically mimicked Scientology, apocalyptic evangelical Christianity, conspiracy theories, and consumerism.[20] It has since gained thousands of adherents, some of whom take the church quite seriously, adapting many of the materials and strategies that made up the '76 Campoon: radio programs, newsletters, books, and annual get-togethers (called "Devivals"). In the 1990s, SubGenius members became some of the earliest and most active participants in Usenet groups on the World Wide Web.

Presiding over and maintaining the Church of the SubGenius is the Reverend Ivan Stang, the fictive clerical persona of Douglass Smith. From the beginning, the chief collaborator of Smith's alter ego Stang has been Dr. Philo Drummond (Steve Wilcox), whose *Puzzling Evidence* program has run for decades on Berkeley's Pacifica station KPFA. At KPFA, Drummond became associated and collaborated with the experimental radio artists Negativland, whose radio show *Over the Edge* began a decades-long run on KPFA in 1981. Inspired by the Situationists, Negativland (through its sensibility), like the Church of the SubGenius (through its direct ties), can also be seen as indirect inheritors of the Firesign ethos, extending the project in their underground work in the years after Firesign lost its connections to mainstream mass media with their own improvisational radio broadcasts and albums of critical sound collages.[21] Negativland's early records were released through their own still extant Seeland Records. In the mid-1980s, their albums began to be distributed by the seminal hard-core label SST (still a far cry from Columbia), through which Negativland released the breakout albums *Escape from Noise* (1987) and its notorious follow-up *Helter Stupid* (1989), the latter a chronicle of the controversy elicited by the *Escape from Noise* track "Christianity Is Stupid," which had included a sample of the Reverend Ivan Stang.

Perhaps the greatest track on *Escape from Noise*, "You Don't Even Live Here," featured contributions from fellow Bay Area artists, the anonymous avant-garde rock band the Residents. Negativland's collaborative ethos and aural world-building itself resembled the work of the Residents, in particular their albums *Eskimo* (1979) and *The Commercial Album* (1980), both of which were released through their own label, Ralph Records. Like Negativland, the Residents have always been more deliberately abrasive than the Firesign Theatre. Like both of those groups, however, they have combined technological experimentation and collective performance in producing a kind of Bakhtinian comedy that was about much more than provoking laughter.

CRATE DIGGING AS ARCHAEOLOGY—FIRESIGN IN HIP HOP

What made possible the "culture-jamming" bricolage of Negativland's late-1980s albums were the newly affordable digital samplers like the Akai S900 and MPC60, the E-MU SP 1200, and the Ensoniq EPS. These same devices were at the same time enabling what would become a far more widely influential revolution, born in the underground, in the world of sample-based hip hop. As Tricia Rose wrote in her foundational 1994 study, the digital sampler is the "quintessential rap production tool," contributing instrumentally to what has come to be known as the genre's late-1980s "golden age."[22] In this context, the Firesign Theatre's Columbia albums began to be given a new and different life—*"alternative lives and alternative meanings"* as Rose describes the fate of the sample—though it is one that seems on the whole to be surprisingly sympathetic with Firesign's project, albeit formed by its own cultural priorities and listening practices.[23] Like the DJs who sampled them, we might say, the Firesign Theatre were critically engaged media artists, deeply conversant with the issues of their time, drawn simultaneously to the technological vanguard and to its historical imbrication. In musicologist Tom Perchard's phrase, hip-hop producers were engaged in a "critical, musical-historical dis-course . . . in which contemporary creative agency was ever intertwined with representations of and relationships to the past, both productive and problematic."[24]

Firesign's Columbia albums have been extensively sampled by many of the most recondite DJs in the history of hip hop, from DJ Mark the 45 King's production on the Chill Rob G track "The Power" (1989) to the Freddie Gibbs, Curren$y, and Alchemist collaboration "Location Remote" on the album *Fetti* (2019) (see appendix B). Taken on its own terms, this discography is testimony both to the aesthetic evolution and to the ethical consistency of sampling practices across three decades. With respect to the latter, across the thirty-odd tracks I have found, there is no sampled phrase that appears more than once (the one exception being Pastor Flash's irre-sistible "I'm high, all right" from *Don't Crush That Dwarf*), and every album other than *The Giant Rat of Sumatra* has been sampled at least once (though I will testify to having heard that one sampled, too, in a trip-hop track at a coffee shop in Philly). This suggests the importance of originality, as Mark the 45 King confirmed in an interview the same year he was pro-ducing Chill Rob G: "People look up to me because I'm looping up records that haven't been used before. I have to buy all the breakbeats records to know what *not* to use."[25] It also suggests the importance not only of know-ing the field but of performing that knowledge through cross-reference.

In what appears to be the first time a record sampled the Firesign Theatre, the 45 King dropped the voice of the animatronic president in *I Think We're All Bozos on This Bus* in a track that took its beat and its title from *Snap!*'s 1989 track "The Power." Two years later, the "War Mix" of Steinski and Mass Media's "It's Up to You"—an up-to-the-minute protest of the first Gulf War—answered by sampling *Don't Crush That Dwarf*'s Deacon E. L. Mouse ("Is it going to be all right?") as the first voice in a devastating tape collage of phrases drawn from the speeches of the then-president, George H. W. Bush. These began with Bush 1's empty apoth-egm on the theme of democratic "power": "We are Americans. Americans know *power belongs in the hands of people*." Steinski and Mass Media's samples cite their original sources, but they are also citations of Mark the 45 King's previous samples: the 45 King sampled Firesign's robot presi-dent in a song called "The Power"; "It's Up to You" samples the Firesign

Theatre before the sitting president's bloviations about the people's "power" as he prepared to declare war.

In these inaugural examples from the late 1980s and early 1990s, the technique is to use a relatively long sampled phrase at the start of a track, playfully or ironically setting the mood. On "Tha Frustrated N——" (1996), DJ Premier cuts up Ossman's lines introducing Nick Danger ("he walks again . . . ruthlessly") as an overture for Jeru the Damaga before the rapper begins his first lyrics; in his 1998 remix of the classic track "Jazzy Sensation" by Afrika Bambaataa and the Jazzy 5 (1981), Steinski samples the courtroom scene in *Don't Crush That Dwarf* ("Youth here doesn't seem to *know about the disappearance of the old school*"), making a pun about old-school hip hop out of the one the Firesign Theatre had intended about square culture's indignation about hip youth culture (on an album where the "old school" of Morse Science literally disappears).

Famous with his partner Double Dee for the early 1980s sample collages "The Lessons," Steinski was by then himself decidedly old school. And at that point, a new hip-hop subgenre of turntablism—which elevated the musical role of the DJ, sometimes without a rapping MC—had begun to flourish both in New York City and on the West Coast in LA and San Francisco. Originally released on cassette by Future Primitive Sound— an important SF space for turntablists including DJ Shadow, Cut Chemist, and Shortkut—the Presage collective's mystically paranoid *Outer Perimeter* record represented another stage of sampling practice in which a single record might be sampled extensively to compose a kind of narrative commentary running over and through the course of an album.[26] In this case Presage's Mr. Dibbs and DJ Jel liberally plunder Firesign's *Everything You Know Is Wrong*, transforming an album that genially spoofs New Age enthusiasms like aliens, alternate histories, and psychokinesis into its obsessional and conspiratorial twin: "Presage is a warning brought to you by Mr. Dibbs, DJ Jel and MC Dose of the 1200 Hobos."

By the twenty-first century, virtuosic producers like J. Dilla and Madlib were embedding shorter, harder-to-identify Firesign samples as more deeply occulted elements within the mix. Madlib, who together with J. Rocc went to meet the Firesign Theatre at one of their last performances, has sampled the Firesign Theatre more than anyone, combining the heteroclite assemblages of West Coast turntablism with golden age skits on albums like *Madvillainy*, a collaboration with MF Doom that included numerous characters and a loose narrative.[27] The historian Seth M. Markle has recently examined Madlib's 2010 album *Beat Konducta in Africa*, understanding its vast array of African and Africa-themed samples as a "'dialogue and commentary' on the historical significance of African and African diaspora music and pan-Africanist struggle." In this eye-opening account, Madlib is presented as "the hip hop DJ as black archaeologist," echoing Tricia Rose's earlier account of sampling as "a means of archival research, a process of musical and cultural archeology."[28] An obvious riff on the practice of "crate-digging," this appearance of the archaeological metaphor is notable because of its importance in a different context in this book, where I have tried to make a case for the Firesign Theatre as practicing a kind of media archaeology *avant la lettre*. But it also raises the question of whether more can be inferred about the place of the Firesign Theatre in hip hop's self-selected archive.

To state the obvious, as monuments of a literary white-stoner comedy, Firesign albums could not have the same significance for Black artists that the African sources forming the 2010 Madlib album would have, to say nothing of the heavily mined and culturally resonant sources like James Brown, Parliament, and 1970s jazz. It would still be possible, nevertheless, to agree with the musicologist Perchard, who respectfully qualifies the historicist argument by suggesting that "while aspects of cultural memory were in play [in hip hop's sampling of jazz recordings], so too were self-interested exploitations of the forgotten and the unknown."[29] Although Steinski, Premier, and the 45 King may have heard the Firesign Theatre as a part of their generation's discography, later DJs like Madlib, J. Dilla, and Alchemist would have had to have discovered them among their parents' collections (Madlib's parents having been musicians themselves) or in the used bins.

There is no single answer to this question, but there is further suggestive evidence in the history of the reception of hip hop at the dawn of its golden age. When De La Soul's first album, *Three Feet High and Rising*, appeared in 1989, it was not only Greg Tate but the *NME*'s Sean O'Hagan and the cultural critic Mark Dery (coiner of the term "Afrofuturism") who compared the group's debut to the Firesign Theatre.[30] This was no doubt in part inspired by the fact that *Three Feet High and Rising* both played with the image of psychedelia and was also very funny, but as Dery insisted, the Firesign Theatre's records were "more semiotic guerrilla warfare than conventional comedy" and De La Soul's debut was "equally surreal."[31] Dery was pointing to the way, anticipating turntablism, the record's huge range of samples (not Firesign but Liberace, Johnny Cash, and Steely Dan alongside the Detroit Emeralds, Barry White, Parliament, and Michael Jackson) itself resembled the enormous storehouse of references making up any five minutes of a Firesign Theatre album (Kent State, the MGM Auction, *Candide*, Buster Keaton, televangelism).

Tate and Dery were also picking up on the phenomenon of the hip-hop skit—a game-show pastiche and other scenarios run through *Three Feet High and Rising*—and the way that a kind of theatrical space might not only sit between songs but also be incorporated as a fictive space within a song. This was an idea that was not unique to De La Soul but could be found throughout near-contemporary recordings by the Pharcyde ("Officer"), Public Enemy ("More News at 11"), Queen Latifah ("Mama Gave Birth to the Soul Children"), and many others. This tradition of diegetic space went back to the presampling era on tracks like the Last Poets' "On the Subway" (1970) which verbally created the space of a New York City subway car setting a dramatic context for Alafia Pudim's poem about a white man's failure to recognize him as they shared a subway car alone. The Last Poets' collectively, verbally produced architecture on this cut resembles the improvised tracks released on the Firesign Theatre's 1972 *Dear Friends* ("Echo Poem," "The Small Animal Administration"), which notably also involved the collaboration of a DJ and engineer, the Live Earl Jive. Combined with the layering effects of multitrack recording (and, in rap music, sampling), these discoveries were each ways of being "two places at once," in the radically realized comic dramas of the Firesign Theatre and the revolutionary musical worlds of golden age hip hop. Is it too much to assume that DJs with ears as big as Madlib's would have intuited a parallel achievement? In Rose's early account, rap music is, in terms sampled from Marshall McLuhan's student Walter J. Ong, a "complex fusion of orality and postmodern technology ... fundamentally literate and deeply technological."[32] In another register, this was the same fusion the Firesign Theatre performed in their own work.

Unsurprisingly, maybe, from across this diverse array of samples from the Firesign Theatre's catalog, there is not one example that includes the Firesign Theatre's adoption of racialized voices (the closest being Wolfman Jack's cameo on Ossman's solo album *How Time Flys*—sampled by both J. Dilla and Madlib—the Wolfman having made his name by appropriating the vocal style of Black DJs for his border radio show in the 1960s). Rather, it seems much more plausible that the DJs who discerningly sampled the Firesign Theatre—whether they were evidently fans or where the attitude is more inscrutable—heard the Firesign Theatre's voices as critical parodies of whiteness (their samples resembling the verbal marking practices of comedians like Rudy Ray Moore and Richard Pryor, who would parody white voices). Placed in the service of a Black music, this recontextualization was an appropriative and interpretive act that may have drawn critical attention to the Firesign Theatre's arrogating aspiration to heteroglossia. Firesign would not have heard their own work as being, in the first instance, a critique of whiteness. But precisely because their own work was critical and parodic, they no doubt would have recognized that reading and approved the purposes their work has since been made to perform.

FORWARD INTO THE PAST

Most if not all of these forms of nondominant culture would be emphatically embraced by the fanzine this book's first chapter began by quoting, *Motorbooty*. Published in Detroit—the historic birthplace of *Creem*, which it acknowledged as a forebear—*Motorbooty* combined satiric and straight takes on its contemporary post-hard-core culture and included lots of comics. It printed approving articles on the extraterrestrial rockabilly artist Von LMO and the Canadian filmmaker Guy Maddin along with appreciative archival pieces on such artists as Blowfly, the Last Poets, the Stooges, the electric Miles Davis, and the Firesign Theatre. Enabled by the explosion of zine culture in the 1990s and by the CD reissues that made old recordings newly available, *Motorbooty* was nevertheless most famous for its elaborate satirical put-ons: a twelve-page article that encyclopedically ridiculed the literary pretentions of rock stars (from Pete Townshend

to G. G. Allin), or the freestanding pull-out *Nuts to You*, which pretended
to be a hard-core punk zine from 1945 ("Fuck paper rationing! Buy two!")
and included corrosive punk-styled capsule reviews of the Andrews Sisters
and Tommy Dorsey, not to mention Benny Goodman's hilariously obscene
and self-regarding tour diary.[33] The Firesign Theatre would have been
right to see themselves sequentially mocked and emulated in these two
pieces.

"Of course we will never listen today as we did then." Those words,
which Trevor Pinch wrote in the quite different context of the Milgram
experiments, could describe the self-authorizing historically engaged
work of the hip-hop artists and in still another register could describe the
snarky archival attitude of the mid-1990s *Motorbooty*. And, as I suggested
at the outset, they have guided my own attempt to reanimate, critically,
the Firesign Theatre. This was also the ethos of the Firesign Theatre's
media archaeology, which found quite different audiences among the DJs
and the *Motorbooty* people, and with Trevor, and with me. The Firesign
records are all still "out there," available as never before, partly on the
streaming services and entirely on YouTube. If they have further afterlives
it will be through conditions wholly transformed from those even of the
1990s. The most fitting homage would be an art that refuses the tempta-
tions of revivalism and the naive presentism that dominates common-
sense reception of media technology and listens, instead, with ears that
point in all directions.

Acknowledgments

First and foundational thanks to my Uncle Nick, without whom all of this would not have been necessary. Thanks, next, to the members of the Firesign Theatre. I was fortunate to meet them together once in the 1990s and still have the note Phil Austin sent me afterwards, encouraging my vague idea about a book. I am sorry neither he nor Peter Bergman have lived to see it appear. Years later, David Ossman and Judith Walcutt welcomed me into their home and patiently endured two weekends of questions. David also allowed me to peruse his astonishing archive, which added immeasurably to this book. He and Phil Proctor tolerated my many queries about long-past minutiae and never interfered with the book as I wrote it (which is what I probably would have done). I hope they recognize their work in this book.

My largest debt is to Firesign Theatre archivist Taylor Jessen. Taylor helped put me in touch with several key figures in Firesign land, provided me rare material, answered an untold number of questions, and graciously read a complete draft of the manuscript before it went to press. My thanks to others in the fan community who connected me with fanzines and airchecks before and in the early days of the internet, Elayne Riggs and Earl Truss in particular. Roger Steffens and John Koethe both made time to talk to me about their experiences listening and sharing material. Jan Pen helped me excavate Peter Bergman's connection to the Amsterdam Provos. Laura Jenemann generously facilitated two crucial days of work at the Library of Congress. The expertise of the Cornell University Librarians has, as always, been invaluable. Thank you Katherine Reagan, Fred Muratori, Simon Ingall, Rhea Green.

It took quite some time to figure out how to write this book. A year's fellowship at the Cornell University Society for the Humanities, as well as the Society's sponsorship of the Cultural Acoustics reading group, gave me space for experiments and false starts. My thanks to Society director Paul Fleming and all the fellows. Early on, Jennifer Stoever provided a much-appreciated forum on her groundbreaking sound studies blog *Sounding Out!* Andrew Leland did the same when he invited me to make a podcast for *The Organist.* I am grateful also for audiences at the Experience Music Project Pop Conference, the Modernist Studies Association, the Louisville Conference on Literature and Culture, the Media Studies Colloquium at Cornell and at Stanford University, UC Berkeley, and UC Davis. My research was materially supported by Cornell University's Milstein Program in Technology and Humanity. Matt Kilbane and Marty Cain provided valuable research assistance, and Luke Dennis helped prepare the manuscript for publication. Ezra Tawil generously offered a space for an eleventh-hour writing retreat; I wish I had been able to thank him for it in person.

This book has benefited hugely from astute readers, most of whom did not know much about its subject. Three readers, most of all, gave their time and read untold numbers of drafts: Damien Keane's comments were sympathetic translations of my discursive drafts and always improved them; Jon Eburne affirmed the book's furthest-flung ambitions, challenging me when I most needed it and encouraging me to write for everybody; Erik Born was the book's media studies bellwether. Debra Rae Cohen, Ingrid Diran, Jane Glaubman, Eric Lott, Tim Morton, Judith Peraino, Sven-Erik Rose, Anna Shechtman, David Suisman, Claudia Verhoeven, and Daniel Williams all read chapter drafts and generously provided comments. Two other readers deserve special mention for reading and for sharing expertise about things I did not know well enough: the anthropologist Brian D. Haley and the Lisp hacker Herb Jellinek. T. Tauri and Dave Q were checks on the DJ questions. And I am grateful for the advice and encouragement of Jim English, Sabine Haenni, Molly Hite, Tom McEnaney, Roger Moseley, Ben Piekut, Nick Salvato, and Gabrielle Civil (who reminded me that it's great to write about something you love).

Raina Polivka saw the book for what it could be, understood it was serious scholarship that needed to be weird, and made a home for it at the University of California Press. Thanks, too, to Sam Warren and Jeff Anderson, who shepherded the book through production and to the readers for the Press, who chose not to remain anonymous: thank you, Jacob Smith and Eric Weisbard.

Gratitude and greetings to my daughters Astrid and Sylvie, who might have time for Firesign but for now seem to prefer Billie Eilish and Wet Leg. No complaints! This book is dedicated to you; let's keep listening to things together. And the most heartfelt thanks, once again, to Rayna Kalas for her gut reaction, for the title before the colon, and for discoveries past and still to come. I love you.

APPENDIX A   Firesign Theatre Discography

1968   *Waiting for the Electrician or Someone Like Him* (Columbia CS 9518); stereo and mono

1969   *How Can You Be in Two Places at Once When You're Not Anywhere at All* (Columbia CS 9884)

         "Forward into the Past" b/w "Station Break" 7" single (Columbia 4-45052)

1970   *Don't Crush That Dwarf, Hand Me the Pliers* (Columbia C 30102)

1971   *I Think We're All Bozos on This Bus* (Columbia C 30737; quadraphonic CQ 30737)

         Dear Friends (twelve-disc set sold to radio stations)

1972   *Dear Friends* (Columbia KG 31099 [2xLP])

         "A Firesign Chat with Papoon" (white label promo Columbia AS41)

         Not Insane (Columbia KC 31585)

1973   Proctor and Bergman, *TV or Not TV* (Columbia KC 32199)

         David Ossman, *How Time Flys* (Columbia KC 32411)

1974   *The Tale of the Giant Rat of Sumatra* (KC 32730)

         Phil Austin, *Roller Maidens from Outer Space* (Epic KE 32489)

Everything You Know Is Wrong (Columbia KC 33141; quadraphonic CQ 33141)

1975   *In the Next World, You're on Your Own* (Columbia PC 33475)

Proctor and Bergman, *What This Country Needs* (Columbia PC 33687)

1976   *Forward into the Past: An Anthology* (Columbia PG 34391 [2xLP])

1977   *Just Folks . . . A Firesign Chat* (Butterfly FLY 001)

1978   Proctor and Bergman, *Give Us a Break* (Mercury SRP-1-3719)

1979   *Nick Danger: The Case of the Missing Shoe* EP (Rhino RNEP 506)

1980   *Fighting Clowns* (Rhino RNLP 018; picture disc RNDP 904)

The Cassette Chronicles (Rhino, six cassettes)

1982   *Lawyer's Hospital* (Rhino RNLP 806)

Shakespeare's Lost Comedie (Rhino RNLP 807)

1984   *Nick Danger and the Three Faces of Al* (Rhino RNLP 8012)

1985   *Eat or Be Eaten* (Mercury 826 452-2 M-1)

1993   *Shoes for Industry! The Best of the Firesign Theatre* (Columbia C2K 52736 2CD)

1994   *Back from the Shadows: 25th Anniversary Reunion Tour* (Mobile Fidelity Sound Lab MFCD 2-747)

1998   *Give Me Immortality or Give Me Death* (Rhino R2 75509)

1999   *Boom Dot Bust* (Rhino R2 75983; DVD 5.1 R9 75979)

2001   *Bride of Firesign* (Rhino R2 74390)

Radio Now Live (Firesign Theatre Records MSUG 102 2CD)

2003   *Alternative Rose Parade* (Firezine 2CDr)

All Things Firesign (Artemis 751167-2)

ARCHIVAL RELEASES

The Firesign Theatre's Pink Hotel Burns Down (1998; Lodestone MSUG 008)

Proctor and Bergman, *Power* (2000; Lodestone)

Papoon for President (2002; Laugh LGH 1130)

The Firesign Theatre's Box of Danger (2008; Shout Factory 826663-10780 4xCD)

Duke of Madness Motors: The Complete "Dear Friends" Radio Era, 1970–1972 (2010; Seeland 534)

Fighting Clowns of Hollywood (2019; Bandcamp)

Drop-Ins (2020; Bandcamp)

Dope Humor of the Seventies (2020; Stand Up SUR 220 [2xLP])

Live at the Magic Mushroom (2021; Not Insane!)

Anytown USA: Live 1974 (2022; Bandcamp)

Fools in Space (2022; Bandcamp)

Before They Changed the Water: Live 1969–1971 (2022; Bandcamp)

Jack Poet Loves You: A Collection of Real Radio Ads, 1968-2000 (2024; Bandcamp)

VIDEO

The Martian Space Party (1972; New Line Cinema)

Proctor and Bergman, *TV or Not TV* (1973; New Line Cinema)

Everything You Know Is Wrong (1974)

J-Men Forever (1979; International Harmony)

Nick Danger in the Case of the Missing Yolk (1983; Pacific Arts)

Eat or Be Eaten (1985; RCA Home Video)

Hot Shorts (1985; PA-85-144 laserdisc)

Weirdly Cool (2001; Rhino Home Video R2 970157)

Everything You Know Is Wrong: The Declassified Firesign Theatre, 1968–1975 (2022; Not Insane! 2xDVD)

APPENDIX B   Firesign Theatre Samples in
Hip Hop and Electronic
Music, 1989–2023

DJ/Producer	Artist	Song	Release	Label and year	Sample
DJ Mark the 45 King	Chill Rob G	"The Power"	*Ride the Rhythm*	Wild Pitch, 1989	"When you clock the human race with a stopwatch, it's a new record every time." (*Bozos*)
Steinski	Steinski and Mass Media	"It's Up to You" (War Mix 1991)	12"	Ninja Tune, 1992	"Is it going to be all right?" (*Dwarf*)
DJ Premier	Jeru the Damaja	"Tha Frustrated N——"	*Wrath of the Math*	Payday, 1996	"Out of the fog, into the smog . . ." (*How Can You Be in Two Places at Once*)
Double Dee and Steinski	Afrika Bambaataa and the Jazzy 5	"Jazzy Sensation (The Jazz Mix)"	12"	Tommy Boy, 1998	"Youth here doesn't seem to know about the disappearance of the old school." (*Dwarf*)
Mr. Dibbs and DJ Jel	Presage	"Divide & Conquer" "Aliens" "Project Lucifer" "The Secret Society" "Remote Control"	*Outer Perimeter* (cassette, then CD)	Future Primitive Sound, 1998	"I don't know how you came by this record . . ." (*Everything You Know Is Wrong [EYKIW]*) "Enough of this deception and trickery . . ." (*EYKIW*) "Seekers everywhere! I was right: everything I knew was wrong!" (*EYKIW*) "Everything you know is wrong. Hello, seekers, here we go again . . ." (*EYKIW*) "Well, is this it? The end of civilization?" (*EYKIW*)
J. Dilla and Madlib	Jaylib (J. Dilla and Madlib)	"The Heist"	*Champion Sound*	Stones Throw, 2003	[police radios]. "It was the whole infernal desert . . ." (*In the Next World*)

Fatboy Slim	Röyskopp	"Eple (Fatboy Slim Remix)"	12"	Wall of Sound, 2003	"Take off your clothes, and come on in." (*Waiting for the Electrician*)
Madlib	Madvillain (MF Doom and Madlib)	"Fancy Clown"	*Madvillainy*	Stones Throw, 2004	"Now subscriber, we've got to punch you up, so please stand by." (*TV or Not TV*)
Madlib	Quasimoto (Madlib and alter ego Lord Quas)	"Civilization Day"	*The Further Adventures of Lord Quas*	Stones Throw, 2005	"Well is this it, the end of civilization? Are we prepared? . . . Classified: ultra secret." (*EYKIW*)
		"Tomorrow Never Knows"			"Could be could be, all I know is everything you know is wrong . . . your father is. . . . Uncle Tom, you're . . .!" (*EYKIW*)
J. Dilla		"Anti-American Graffiti"	*Donuts*	Stones Throw, 2006	"Have mercy have mercy, you bet your life, baby. I'm down here at the world's most popular drinking area, you understand: the South Pole Lounge, where everything is nice and they do it twice." (*How Time Flys*)
Madlib	Madlib the Beat Konducta	"Offbeat (Groove)"	*Vol. 1–2 Movie Scenes*	Stones Throw, 2006	"Now please everyone, lock your wigs, let the air out of your shoes and prepare yourselves for a period of simulated exhilaration." (*Bozos*)
Madlib	Talib Kweli and Madlib	"Time Is Right"	*Liberation*	Blacksmith Music, 2007	"Well, I'm impressed most of all with the serene mood, the lack of comprehension." (*How Time Flys*)
Madlib	Madvillain	"Fire in the Hole"	*Madvillainy 2: The Madlib Remix*	Stones Throw, 2008	"I'm high on the real thing . . . over . . . I'm high all right." (*Dwarf*)

DJ/Producer	Artist	Song	Release	Label and year	Sample
Mister Modo and Ugly Mac Beer	Mister Modo and Ugly Mac Beer	"Danger Modo"	*Modonut*	KIF, 2008	"Hello, dear friends. . . . You must be way out there. . . . I'm high all right but not on false drugs." (*Dwarf*)
Madlib	Madlib the Beat (feat. Roc C & Oh No)	"Take That Money"	*WLIB: King of the Wigflip*	BBE, Rapster 2008	"So here it is, at the Saturday Night Gun Mart, a double barrel, over and under pump spray, fully automatic . . ." (*Give Us a Break*)
Madlib	Madlib (feat. J. Dilla and Guilty Simpson)	"Young Guns"	*Medicine Show #1: Before the Verdict with Guilty Simpson*	Madlib Invazion, 2010	"Sorry to interrupt, but . . ." (*How Time Flys*)
		"My Moment (OJ Simpson Remix)"			"This sounds like the end of the world all over again, man. Right on, right on." (*How Time Flys*)
Madlib	Guilty Simpson	"Prelude"	*OJ Simpson*	Stones Throw, 2010	"Is this on? Are we on? (*Roller Maidens*)
		"Pimp Rap Interlude"			"Yes, it's good to see you all again. . . . That's right, I'm Gilbert Skink, and did you know that poison oak makes an excellent tea?" (*Roller Maidens*)
		"Scratch Warning"			"Are they through, are they finished now? Have they played all of their tunes? . . . Don't forget if all you can do is sit at home, then try to keep yourselves awake." (*Roller Maidens*)
		"Hold Your Applause Interlude"			"I want you to hold down your applause for as long as you can because . . ." (*Roller Maidens*)

Artist	Producer	Track	Album	Label, Year	Citation
		"Trendsetters"			"You gotta get over here, right away . . . thrilling end of tonight's two-part episode . . . Paranoid Broadcasting System." (*Roller Maidens*)
		"Outro"			"Here comes Ritardo. Goodbye. Goodbye. Goodbye." (*Roller Maidens*) "Sorry to int—" (*How Time Flys*)
J. Dilla and Madlib	Madlib	"Louder (Blast Your Radio Theme)"	*Medicine Show #11: Low Budget High Fi Music*	Madlib Invazion, 2011	" . . . coming into consciousness now. And you just want to keep still while you warm up. . . ." (*How Time Flys*)
C. Scott	C. Scott	"Don't Crush That Dwarf"	*Stage Theory (Beats, vol. 3)*	2012	Title citation
Wylie Cable	Wylie Cable		*How Can You Be in Two Places at Once When You're Not Anywhere at All?*	Dome of Doom, 2013	Title citation
Madlib	Madlib	"Tarot Ash"	*Rock Konducta*	Madlib Invasion, 2014	"Doors open in . . . comfy." "I like the future . . . it's electric." (*Bozos*)
Alchemist	Freddie Gibbs, Curren$y, Alchemist	"Location Remote"	*Fetti*	ESGN / ALC Laboratories / Jet Life Recordings, 2019	"That is just exactly what they want you to believe." (*EYKIW*)

DJ/Producer	Artist	Song	Release	Label and year	Sample	Title citation
The Orb	The Orb	"why can you be in two places at once, when you can't be anywhere at all (where's gary mix)"	*Prism*	Cooking Vinyl, 2023		

Notes

PREFACE

1. The punk band was called Shoes for Industry; the issue of *Defenders* is #34, "I Think We're All Bozos in This Book"; legendary fan Ashbery quotes Firesign in "Text Trek," *Commotion of the Birds* (32–33); the dynamite recipe from Mrs. Tirebiter appears on page 13 of the January 13, 1973, issue of the *Tribune*; Howard Roberts's 1971 album *Antelope Freeway* both cites and emulates Firesign; Greg Tate, "Yabba Dabba Doo-Wop: De La Soul," in *Flyboy in the Buttermilk*, 140; Lennon's "Not Insane" pin is discussed in chapter 2; Madlib's tracks are listed in appendix B and discussed in the coda.

2. Garson, "Luddites in Lordstown," 72.

3. Scoppa, "Los Lobos."

4. Toubkin and Vellender, "Lone Survivor's Guide to Firesign Theatre," 13.

5. Situationist International, *Ten Days That Shook the University*.

6. McLuhan, *Understanding Media*, 23.

7. Christgau, "Consumer Guide," Jan. 17, 1995.

8. Pinch, *Analog Days*.

9. Pinch, "Stanley Milgram and the Sonic Imaginary."

LINER NOTES

1. Wiebel, *Backwards into the Future*, 251.

2. Ossman, *The Sullen Art*. Recently, Lisa Hollenbach has said Ossman's radio show "functioned in a sense as a kind of radio supplement to *The New American Poetry* [Donald Allen's seminal 1961 anthology]." She has been critical of both Allen's and Ossman's neglect of poets of color and the movement's "compulsory homosociality." Hollenbach, *Poetry FM*, 128–30.

3. Ossman, "A Memoir of the Renaissance Pleasure Faire."

4. Glass, *Counterculture Colophon*, 65–99.

5. Esslin, 327–36, 408.

6. Ghelderode, "Christopher Columbus," 174–75.

7. Black, *The Tornado in My Mouth*; *KPFK Folio*, March 1967.

8. Wiebel, 89–92, 100; Blau, *Programming Theater History*.

9. Trina Robbins, *Last Girl Standing*, 45–58; Lambert, *Good Vibrations*, 174–76.

10. Dallas, *Dallas in Wonderland*, 36–61.

11. United States [US], Dept. of Justice, Peter Bergman file.

12. Hayden and Bergman each remembered that the recruitment was overseen by Gloria Steinem. Hayden remembers:

> In 1962, curious about these youth festivals and eager to see the world, I interviewed as a possible participant in an American (anti-communist) delegation to the Soviet-sponsored Helsinki Youth Festival in Finland, one of several of the era. . . . Personally, I never made it into a CIA front group, though I tried hard enough. I was "unwitting," in spook-speak. "Witting" was what the agency called people in the know. They first tested and recruited them into high positions in the student world, then administered a surprise security oath before telling them they were part of the CIA.
>
> (Hayden, "The CIA's Student-Activism Phase"; see also Paget, *Patriotic Betrayal*)

13. Except where otherwise noted, details of Bergman's biography are drawn from his interviews with Maryedith Burrell, recorded for an unrealized memoir in 2003–4.

14. Lee, *Tom Stoppard*, 96–100. On the relation of the Ford Foundation to the CIA and US Cold War initiatives, see Berghahn, *America and the Intellectual Cold Wars in Europe*, 214–49.

15. Biner, *The Living Theatre*, 84–102; Tytell, *The Living Theatre*, 197–211. Artaud's *The Theater and Its Double* opens with an essay that analogizes the condition of the theater to that of the plague, as described in Boccaccio's *Decameron*. "First of all we must recognize that the theater, like the plague, is a delirium and is communicative. . . . [Yet] if the essential theater is like the plague, it is not because it is contagious, but because like the plague it is the revelation, the bringing forth, the exteriorization of a depth of latent cruelty by means of which all the perverse possibilities of the mind, whether of an individual or a people, are localized." Artaud, *The Theater and Its Double*, 27, 30.

16. On the buildup of arts initiatives and soft power in Berlin circa 1964, see Saunders, *The Cultural Cold War*, 351–53.

17. Stern, "A Short Account of International Student Politics," 30–32. See also Saunders 382–83; and Berghahn 244–49.

18. "Dit Nummer is een Happening," *Ratio.*

19. Reeuwijk, *Damsterdamse Extremisten*, 36–43.

20. Pas, "Mediatization of the Provos," 159, 169–71.

21. Proctor and Schreiber, *Where's My Fortune Cookie?*, 53–62.

22. Fong-Torres, *Hickory Wind*, 60–73; Proctor, "The History of Firesign, Part 2."

23. Savage, *1966*, 480–83.

24. Mike Davis, "Riot Nights on Sunset Strip," 313, 321.

25. "Actor Peter Fonda Is Taken into Custody," 10.

26. "Provos: Blasting Back at the Brave New World," 3; Wiebel, *Backwards into the Future*, 160–61.

27. McCarthy, *On the Appearance of the Comedy LP*, 23–24.

CHAPTER 1

1. Merline, "The Firesign Theatre," n.p.

2. Lask, "Comedians," D 23.

3. Merline, "The Firesign Theatre," n.p.

4. HBO broadcast its first hour-long stand-up comedy special at the end of 1975, the year of the Firesign Theatre's final Columbia album. Silicon CDs outsold vinyl albums for the first time in 1987, the same year Firesign's early albums began to be reissued as audiophile CDs by Mobile Fidelity Sound Lab.

5. The biggest jump occurred in 1968 ($228 million). Recording Industry Association of America information reproduced in Marsh and Stein, *The Book of Rock Lists*, 19–21; see also Chapple and Garofalo, *Rock 'n' Roll Is Here to Pay*, 76.

6. Christgau, "New Kind of Comedy Served on a Platter."

7. Duncan, rev. of . . . *Is It Something I Said?*, 84.

8. Christgau, "New Kind of Comedy Served on a Platter." Other acts unevenly following the Firesign Theatre's path included Cheech and Chong, the Credibility Gap, Conception Corporation, Hevy Gunz Industries, and Congress of Wonders. Ian Brodie has recently suggested that Cosby's late 1960s monologues began to accommodate the album form with increasingly discursive and jazz-like riffs. Ian Brodie, *A Vulgar Art*, 188–95.

9. Christgau, "Consumer Guide," Dec. 26, 2000.

10. Sterne, *The Audible Past*, 182.

11. Gelatt, *The Fabulous Phonograph*, 187.

12. Horning, *Chasing Sound*; Anderson, *Making Easy Listening.*

13. Wilfrid Mellers, quoted in Frith, *Sound Effects*, 50.

14. Willis, "Rock, Etc.: Pop Ecumenicism," 175–76. Drugs themselves, as "extensions of the senses," have been taken to be a kind of media by theorist-provocateurs such as Marshall McLuhan and Friedrich Kittler.

15. Tate, "Yabba Dabba Doo-Wop," *Flyboy in the Buttermilk*, 137–38, 140.

16. Richard Hill, "Chanting the Square Deific," *Rolling Stone*, Sept. 30, 1970, 538.

17. "Four Hours of Firesign Theatre," *Cornell Daily Sun*, 7.

18. Verma, *Theater of the Mind*, 17–32.

19. Marcus, "The Firesign Theatre," 130.

20. Bakhtin, *The Dialogic Imagination*, 49. Subsequent citations of pagination from this source are given parenthetically in the text.

21. Bangs, "Not Insane. And Not Funny Either," 68.

22. Jennifer Lynn Stoever coins the term "aural blackface" in *The Sonic Color Line*, 8–11.

23. Tate, *Everything but the Burden*. For an example of the condescension, see the Credibility Gap's "An Evening with Sly Stone," on 1973's *A Great Gift Idea*.

24. *Oxford English Dictionary*, s.v. "stratification (n.)," 4: "The process or result of being formed or arranged into layers; the fact or state of having layers. Also *figurative* with reference to things conceived of as constituting a series of layers. Frequently in archaeological contexts; . . . 6. The formation or establishment of social or cultural strata resulting from differences in occupation and political, ethnic, or economic influence. Cf. *social stratification* n."

25. Ernst, "Media Archaeology," 239; Elsaesser, "Media Archaeology as Symptom," 181–215.

26. Tom Perrin notes that the phrase "Great American Novel" appeared in the *New York Times* vastly more times in the 1970s than in any other decade and suggests that the decade's novels by John Updike, Saul Bellow, Eudora Welty, Vladimir Nabokov, Toni Morrison, Thomas Pynchon, Ishmael Reed, and E. L. Doctorow can all be seen to engage the concept, in addition to books by Philip Roth and James Fritzhand, which are explicitly (and ironically) titled "Great American Novel." Perrin, "The Great American Novel in the 1970s," 196.

27. Marsh, rev. of *I Think We're All Bozos on This Bus*, 60.

28. Ossman, *Dr. Firesign's Follies*, 17.

29. Eisenberg, *The Recording Angel*; Ashby, *Absolute Music, Mechanical Reproduction*, 4. See also Ray, "Tracking."

30. Glass, *Counterculture Colophon*, 30.

31. Mills, "The Firesign Theatre," *Fusion*, 16; Edmonds, "The Velvet Underground," *Fusion*, 20.

32. Hunt, *The Textuality of Soulwork*, 1; Lesnick, *Guerilla Street Theater*; Artaud, *The Theater and Its Double*, 74–83.

33. Ossman, "A Memoir of the Renaissance Pleasure Faire."

34. Malloch, interview by Chris Palladino, June 1, 1995.

35. Haley, *Hopis and the Counterculture*. See also Haley, "Craig Carpenter and the Neo-Indians of LONAI."

36. Garson, *Macbird!: A Recording of the Complete Text of the Play with the Original Cast*.

37. Robert Christgau would devote almost an entire page to eviscerating *The Astrology Album* in the first issue of the short-lived rock journal *Cheetah*. Christgau, "Records," 18, 21.

38. Gendron, *Between Montmartre and the Mudd Club*, 185–86.

39. Ellen Willis, "Records: Rock, Etc.," *New Yorker*, August 10, 1968, 87; Christgau, *Any Old Way You Choose It*, 1.

40. Keightley, "Long-Play"; Anderson, *Making Easy Listening*.

41. Turner, *Beatles '66*, 56–57, 102, 119.

42. "Comedy and Spoken Word Spotlights—1961," 18.

43. Wiebel, *Backwards into the Future*, 45–46, 98–100.

44. Freberg, *It Only Hurts When I Laugh*, 201–13.

45. Zinn, *A People's History of the United States*, 1.

46. Verma, *Theater of the Mind*, 35–38.

47. Haley, *Hopis and the Counterculture*; Deloria, *Playing Indian*, 86; Vizenor, *Manifest Manners*, 3–5. A more sympathetic view of the dynamic can be found in Smith, *Hippies, Indians, and the Fight for Red Power*.

48. All quoted material in this paragraph comes from Haley, *Hopis and the Counterculture*.

49. Haley, *Hopis and the Counterculture*.

50. Haley.

51. Bergman, "Peter Bergman on the Hopi Indian," 8.

52. Hochman, *Savage Preservation*.

53. Firesign Theatre, "A Shadow Moves upon a Land," *Profiles in Barbecue Sauce*, 48. A recording of the performance is included on *The Firesign Theatre at the Magic Mushroom*.

54. Ossman, *Dr. Firesign's Follies*, 18–19.

55. Maybe inevitably, Close's record was itself ripping off a book: Cab Calloway's *Cat-ologue: A Hepster's Dictionary* and its instructional follow-up *Prof. Cab Calloway's Swingformation Bureau*.

56. Echoes of Havel's *The Garden Party*, which involved the attempt to abolish the Department of Liquidation, can be heard on the Firesign's 1970 album *Don't Crush That Dwarf, Hand Me the Pliers* as Principal Poop hails the "Department of Redundancy Department."

57. Kate Steinitz, interview by Philip Proctor and David Ossman, summer 1967.

58. Glass, *Counterculture Colophon*, 95.

59. Gendron, *Between Montmartre and the Mudd Club*, 161.

60. Gendron, 29, 73.

61. Ed Ward, rev. of *Waiting for the Electrician or Someone Like Him*, 56.

62. Howard, *Sonic Alchemy*, 70.

63. Rogan, *The Byrds*, 197.

64. Both singles were sufficiently far out that Clive Davis required them to be remixed as conventional songs when they were released on the Sagittarius LP *Present Tense* in 1968. Lenny Kaye, however, chose the original seven-inch mix when he included the song on the influential *Nuggets* compilation in 1972.

65. Kubernik and Kubernik, *A Perfect Haze*, 17.

66. Ossman, interview by the author, April 14, 2016.

67. Phil Austin explained Firesign's unique arrangement:

> Because we were not a musical act, we did not receive publishing royalties the way a musician does, what are called mechanicals. So Columbia compensated us by raising our royalty rate up to approximately the level of Frank Sinatra, which still didn't give us enough compensation, but one of the things they threw in was unlimited studio time. So when we were in the studio, we were not charged for it essentially. We developed our writing in a way with having unlimited studio time. Our engineers would work with us till five in the morning. We would stop in the middle of something and say, "Gee, that isn't right, so let's go rewrite it." Then we would be out of the studio for five days and then we would come back in, and work and work, and stop and go, and stop and go.
>
> (Quoted in Wiebel, *Backwards to the Future*, 136)

68. Wald, *How The Beatles Destroyed Rock 'n' Roll*, 236. Subsequent citations of pagination from this source are given parenthetically in the text.

69. Sterne, "The Stereophonic Spaces of Soundscape," 69, 73.

70. Weheliye, *Phonographies*, 127 (my italics).

71. Piekut, *Henry Cow*, 394, 396.

72. Chartier, "The Practical Impact of Writing," 118. Subsequent citations of pagination from this source are given parenthetically in the text.

73. Stadler, "'My Wife,'" 426.

74. Chartier, "Labourers and Voyagers," 90.

75. Keightley, "Long-Play," 380–82.

76. *Rolling Stone*, Sept. 17, 1970; *Creem*, March 1970.

77. Tate, "The Electric Miles (Part 1)," *Flyboy in the Buttermilk*, 69–70.

78. Tingen, *Miles Beyond*, 41.

79. Tingen, 43–48.

80. Tingen, 72.

81. Tomlinson, "Cultural Dialogics and Jazz," 92–93, 102.

82. Bakhtin, 82. In *Scripts of Blackness*, Noémie Ndiaye observes instances of early modern commedia dell'arte that are defined by their use of racialized voices, or what she terms "blackspeak" (87–88, 169–70).

83. Elsaesser, "Media Archaeology as Symptom," 196.

84. Rubin, *Well Met*, 14.

85. Rubin, 21.

86. Ossman, "A Memoir of the Renaissance Pleasure Faire."

87. Wilcock, "A Love-In Inventory," 1.

88. Turner, *The Democratic Surround*, 291. Subsequent citations of pagination from this source are given parenthetically in the text.

89. The phrase "secondary orality" was coined by McLuhan's student Walter J. Ong. See Ong, *Orality and Literacy*, 1–15.

CHAPTER 2

1. Firesign Theatre, *Radio Hour Hour*, Feb. 1, 1970, *Duke of Madness Motors*, DVD.

2. Kael, "Bonnie and Clyde," 148.

3. Ossman, interview by Kurt Ericson, May 7, 2015.

4. McLeod, *Pranksters*.

5. Barnouw, *The Golden Web*, 79–89.

6. McKenzie, *Bibliography and the Sociology of Texts*, 93.

7. Ward, "Postmark: Deep Space," 3.

8. Cull, *The Cold War and the United States Information Agency*.

9. Former USIA director Alvin Snyder provides an account of the Costa Rica operation in *Warriors of Disinformation*, 193, 199–200.

10. Cantril, Gaudet, and Herzog, *The Invasion from Mars*.

11. Everett, "Firesign Theatre."

12. Kael, "Bonnie and Clyde," 150.

13. Taylor, "7 Times the Onion was Lost in Translation."

14. Peck, "The Column," 9.

15. McLuhan and Fiore, *War and Peace in the Global Village*, 134.

16. Finkelstein, *Sense & Nonsense of McLuhan*, 117.

17. Roberts, *The Psychological War for Vietnam*, 10.

18. Austin and Peter Bergman were both still enlisted when *How Can You Be* was released. Austin would remember, "Psychological Warfare had to do with radio, that's really what it was really. We were supposed to be set up to be sent to Vietnam and broadcast to the Heathen about what a great life we were going to have for them." Austin's commanding officer was Noel Blanc, son of Looney Tunes' legendary "man of a thousand voices," Mel Blanc. Wiebel, *Backwards into the Future*, 93–94, 110.

19. He was later a *Jeopardy!* champion. Stockwell, "Former Breakfast Special Host Chris Ward Passes."

20. Ossman, liner notes for *Waiting for the Electrician or Someone Like Him*.

21. Steffens, interview by the author, May 31, 2017.

22. Perry, "Is This Any Way to Run the Army—Stoned?," 6.

23. On Hammond's support for the Firesign Theatre, see Wiebel, *Backwards into the Future*, 127; on McClure's support, see Carpenter, "Dr. Firesign's Travelling Antique Circus," 36, 30; and the September 1970 *Fusion* magazine interview on *Duke of Madness Motors*.

24. Ossman, personal communication, March 25, 2019.

25. "Top LP's," Oct. 18, 1969, 82–84.

26. Christgau, "Consumer Guide," Dec. 11, 1969.

27. Christgau, *Any Old Way You Choose It*, 130.

28. Gendron, *Between Montmartre and the Mudd Club*, 189–224.

29. Barry, "High-Fidelity Sound as Spectacle and Sublime, 1950–1961," 120; Eno, "The Studio as Compositional Tool," 186–87.

30. Barry, "High-Fidelity Sound as Spectacle," 117.

31. McKinney, *Magic Circles*, 183.

32. Blesser and Salter, *Spaces Speak, Are You Listening?*, 163.

33. Blesser and Salter, 165.

34. "From the 1960s . . . the car radio was increasingly presented as a device that could help motorists tolerate their fellow traffic participants and keep them, emotionally, *at a distance*. . . . Such intimacy had also been evoked by car makers since the mid-1940s, presenting the car as a living room on wheels. But in the 1960s, the car radio was added as a personal mood regulator. Those who had a car radio would be able to cope with the outside world and 'reconcile' themselves with it." Bijsterveld, "Acoustic Cocooning," 198.

35. Granata, *I Just Wasn't Made for These Times*, 133–35.

36. McGuinn and Crosby "'Interview' for DJs."

37. "New Recording Facilities for Technical Operations."

38. Verma, *Theater of the Mind*, 41–42.

39. Originally *The Mercury Theatre on the Air*, *The Campbell Playhouse* began broadcasting in New York and concluded its final seasons in Hollywood, Welles's relocation to Los Angeles (and movies) occurring around the time Columbia Square opened. Higham, *Orson Welles*, 140–41.

40. Corwin quoted in Linder, "A Salute to Columbia Square."

41. The one exception was the "Station Break/Forward into the Past" single from 1969, which used condenser microphones. Ossman, interview by the author, April 14, 2016.

42. Verma, *Theater of the Mind*, 99.

43. Verma has written that American radio of the 1930s routinely "destabilized the place of listening." Because its narratives "wrestled with time and space in every available way," radio's audience had become acculturated to the death of "the old Euclidian god." That this should have happened ahead of the age of high fidelity is notable to read in relation to the commentaries of Blesser and Salter and Eno, and it suggests that radio fiction, then only a decade gone, is an

unacknowledged forebear of Stockhausen, psychedelia, and multitrack "in-studio composition." *Theater of the Mind*, 48, 28, 25.

44. Wiebel, *Backwards into the Future*, 162.

45. Elsaesser, "Media Archaeology as Symptom," 183–84. Subsequent citations of pagination from this source are given parenthetically in the text.

46. *Manual of German Radio*, quoted in Schafer, "The Music of the Environment," 37.

47. Thompson, "The Great War of Words," 69.

48. Keane, "An Ear toward Security."

49. Horten, *Radio Goes to War*, 3.

50. Norman Corwin, *On a Note of Triumph*, 9, 11.

51. Ossman, "The Odyssey of Me and Norman Corwin," 216.

52. Verma, *Theater of the Mind*, 81.

53. Script notes for *HCYB* indicate that "This land is made of mountains" is to be sung in the style of "*PETE SEEGER* AND CHORUS." Pete Seeger and the Almanac Singers' song "Rally 'Round Hitler's Grave" serves as a leitmotif in Corwin's *On a Note of Triumph*.

54. Terry, *Bloods*, xv–xvi.

55. Ossman, "The Odyssey of Me and Norman Corwin," 214.

56. Kittler, *Gramophone, Film, Typewriter*, 96–97. Subsequent citations of pagination from this source are given parenthetically in the text. This section of the chapter takes its heading from Winthrop-Young, "Drill and Distraction in the Yellow Submarine."

57. As Edgar Bullington recalled, "He did produce it just like music." Edgar Bullington and Glen Banks, interview by the author, May 23, 2017.

58. Historian Meredith Lair has dubbed this the "total war on boredom"; Michael Kramer calls it "hip militarism." Lair, *Armed with Abundance*, 158–60; Kramer, *Republic of Rock*.

59. According to Andrew Hickey, Hendrix was emulating the Small Faces' "Itchycoo Park." Hickey, "'Itchycoo Park,' by the Small Faces."

60. Corwin, "We Hold These Truths," 86–87.

61. Geoff Emerick, quoted in Lewisohn, *The Beatles Recording Sessions*, 81.

62. Christgau, "Consumer Guide," Jan. 17, 1995.

63. Lennon, "The Goon Show Scripts," G6.

64. Gould, *Can't Buy Me Love*, 49–50. Subsequent citations of pagination from this source are given parenthetically in the text.

65. Lennon, "The Goon Show Scripts," G6. Richard Lester, who directed *A Hard Day's Night*, had worked with Milligan and Sellers on *The Running Jumping Standing Still Film* (1959). The flexi-disc records the Beatles made each Christmas for their fan club are obviously recorded under the sign of the Goons and by 1966 uncannily begin to resemble Firesign Theatre albums.

66. Ossman and Austin, quoted in Ventham, *Spike Milligan*, 192–94.

67. Milligan, *Adolf Hitler: My Part in His Downfall*.

68. See also Jem Roberts's description: "Each week's story, heavily laden with mind-stretching developments conveyed to the fans via ever-more ambitious sound effects, was almost invariably obsessively rooted in Milligan's wartime experiences—making the show a sustained attack on authority and pre-War ideas of respect for the establishment . . . instilling a social rebellion in British youth like nothing which came before it." Roberts, *Fab Fools*, 11.

69. Emerick and Massey, *Here, There, and Everywhere*, 122–23.

70. Quoted in Marcus, "The Beatles," 183.

71. Goldstein, "We Still Need the Beatles, but . . . ," 98–99.

72. Quoted in Sheffield, *Dreaming the Beatles*, 144.

73. "It was joyful rhythm, generosity, youthfulness and communality of their voices. They were four guys who were a gang, they loved and appreciated each other. . . . The Beatles changed American consciousness—introduced a new note of complete masculinity allied with complete tenderness and vulnerability." Ginsberg, "Beatles Essay," 456.

74. Eldon, foreword to Roberts, n.p.

75. Edgar Bullington and Glen Banks, fans who were invited to be in the room during the radio shows of 1971 and 1972, recall:

BANKS: At the very beginning of the show they'd do a little prayer, a little

BULLINGTON: Mantra?

GB: They would hold hands. They would stand in a circle and hold hands and put their heads down I believe

EB: That sounds familiar.

GB: And they would kinda go like this . . . and I don't remember if they said anything, but they held hands and they had like a connection before show time.

Interview by the author, May 23, 2017

76. John Ashbery's obsessive Firesign Theatre fandom began the evening his friend Peter Delacorte recited one side of *How Can You Be in Two Places at Once* from memory. Koethe, "Ninety-Fifth Street," 372. See also Greathouse, "Firesign Fanatics Freak-out Tonight," 4.

77. Marcus, "Rock-A-Hula Clarified," 39–40.

78. Christian [Marcus and Miroff], "The Masked Marauders," 392–94. See also McKinney, *Magic Circles*, 269–72.

79. Marcus, "The Beatles," 181.

80. Reeve, *Turn Me On, Dead Man*, 124–25, 143.

81. Bromell, *Tomorrow Never Knows*, 14.

82. Robert A. Ducksworth-Ford, "Column: Is Paul Dead? or Are You?," *Los Angeles Free Press*, Nov. 7–13, 1969, 40.

83. Speaking of islands, compare this Q drop: "Welcome to Epstein Island. Ask yourself, is this normal? What does a 'Temple' typically symbolize? What does an 'OWL' symbolize (dark religion)? Tunnels underneath? Symbolism will be their downfall. These people are EVIL." https://www.reddit.com/r /CoincidenceTheorist/comments/cbnx3a/welcome_to_epstein_island_ask_ yourself_is_this/.

84. Christgau, "Album of the Year."

85. "Manson Is Music 'Addict,'" 17.

86. Belz, *The Story of Rock*, 212.

87. Le Blanc and Davis, *5 to Die*, 124. The December 19, 1969, issue of *Life*, with its famous cover story on Manson, mentioned the Beatles connection in passing; *5 to Die* was the first publication to elaborate the theory through readings of Beatles songs. My thanks here, and in all things Manson, to Claudia Verhoeven.

88. Christian, "Masked Marauders," 389–90.

89. Walls, rev. of *How Can You Be in Two Places at Once When You're Not Anywhere at All*, 26. This same issue of *Creem* contained Deday LaRene's "The Walrus Was Paul?" (5–8), an article that professed to debunk the Paul-is-dead rumor but exhaustively included all the Beatles' "clues," and thereby amplified it.

90. Hughes, *Chasing Shadows*.

91. Gosling, *Waging "The War of the Worlds,"* 130–42. Both of the WKBW broadcasts are stored at the Internet Archive, https://archive.org/details /wkbw-halloween-archive.

92. Simpson, *Science of Coercion*, 39.

93. Gosling, *Waging "The War of the Worlds,"* 137–38.

94. Neer, *FM*, 77; Pierce, *Riding on the Ether Express*, 321; Nisker, *If You Don't Like the News . . .*, 49–50.

95. Krieger, *Hip Capitalism*, 109.

96. Walker, *Rebels on the Air*, 72–74. Walker points out that Fass's eclecticism and charisma drew in part from earlier examples such as Jean Shepherd and Black AM DJs of the 1950s.

97. Likewise, its circular construction—signaled by Ralph Spoilsport's pot-selling reprise—was a particularly clear riff on the freeform aesthetic, DJs often using the final track of a long sweep of sounds to tie the set together. And the fact the group tracked the piece in order from beginning to end, yet commenced recording before knowing how it would resolve, meant that the connection to freeform's improvisational sequencing was practical as well as thematic.

98. Hayes, "KRLA Airchecks." According to Walker, Fass had once collaged "a Hitler speech with a Buddhist chant in the background" (*Rebels on the Air*, 73).

99. "Peter Bergman's 'Oz' . . . Happening," 12.

100. "Our model was *The Goon Show* and we knew because I had interviewed people from *The Goon Show* and knew how the show was done. Like Sellers

would make this big entrance and there was pre-show stuff, and just also the unexpected funny voices or funny characterization, beyond the funny *writing*, you know. And we all could come up with one of those, as we learned. Because we were learning how to write scripts, really" (Ossman, interview, April 14, 2016).

101. Keith, *Voices in the Purple Haze*, 29.

102. See Krieger's classic study of KMPX and KSAN, *Hip Capitalism*.

103. Simpson, *Early '70s Radio*, 91–124.

104. Bergman was fired from KRLA for extemporizing ad copy on the air ("Get your hand up the skirt of a Toyota and you'll never let go!") and from KMET for playing a forbidden Fugs song, "Johnny Pissoff Meets the Red Angel." Jessen, "An Avant-Garde Religious Radio Happening," 14.

105. "Programming Aids," 50–52.

106. WABX DJs Larry Monroe and Jim Dulzo, quoted in Fenton, "'We're Selling Too Much Death,'" 7.

107. Torey, "Melted Cheese," 4; Harrington, "WHFS: The End of the Rainbow," G6.

108. Reeve, *Turn Me On, Dead Man*, 107.

109. Gold, "Muzak of the Spheres," Nov. 13–31, 1972, 22–23. See also "'Obscene' airwaves," 4.

110. Richard, "'Cheese' Melts Off Air," C6.

111. Krieger, *Hip Capitalism*, 150–58, 166–70. Fong-Torres, "FM Underground Radio: Love for Sale," 1, 6, 8.

112. Chapple and Garofalo, *Rock 'n' Roll Is Here to Pay*, 114–15. In October 1971, the entire KPPC staff was fired.

113. Barrett, "Lew Irwin Sets the Record Straight."

114. Bergman described it in 1970: "It was to be a live show at this big initiation lodge, it was a *huge* place, and we were going to have interviews and an audience and a theatre organ and Donovan was going to come on." Firesign Theatre, *Fusion* magazine interview, Sept. 1970, *Duke of Madness Motors*, DVD.

115. Austin has said that *Yours Truly, Johnny Dollar* was the explicit model, but this narratological feature, as well as the way "The Further Adventures of Nick Danger" echoes *The Adventures of Philip Marlowe*, suggests that the latter is the more direct antecedent. Furthermore, one of Nick Danger's most famous gags is prefigured directly by the beginning of *Philip Marlowe*'s episode "The Red Wind": "I closed up my office early. I got tired of reading 'Philip Marlowe, Private Investigator' backwards on the ground glass of my office door" (air date Sept. 26, 1948).

116. Kittler, "Playback," 103–6.

117. Barnouw, *The Golden Web*, 79–83.

118. Thompson, "On the Record," 21. Another journalist who witnessed the Putsch wrote, "The important fact, however, is that in this attempted revolution the broadcasting headquarters was deemed a primary point of attack—and so it

will be in all revolutions from now on." Saerchinger, *Hello America!*, 204. See also Gunther, *Inside Europe*, 291–308.

119. Krieger, *Hip Capitalism*, 149.

120. Hexagram 7 (Discipline, or "The Army"); line 6 at the top. *The I Ching, or Book of Changes*, 36.

121. They may also have known that using the *I Ching* as both reference and compositional tool to compose a counterfactual narrative describes precisely the plot of a 1962 novel speculatively set in a United States that had lost World War II, Philip K. Dick's *The Man in the High Castle*.

122. Phil Austin discussed this at length in 1988:

> Most important for us, obviously, is the reference at the end of the story to the "I Ching," that ancient Chinese divining tool, composed of hexagrams of six lines, each capable of a dualistic reversal to its opposite. We vowed that we would follow the instructions for the casting of the Oracle to the letter and that we would faithfully record the results at the end of the story of Nick Danger. We would trust to the falling pocket-change of Chance. . . . We threw "The Army" with the changing line in a position to lead to the hexagram "Youthful Folly." It was a minor miracle—or at least an affirmation from the Ether that we were on some right track or other. Nick, out of the Army, becomes Babe/Bill, the Youthful Fool. Nick surrenders. The war is over. . . . All we had to do now was get through 1969.
>
> (Austin, liner notes to *How Can You Be in Two Places at Once When You're Not Anywhere at All*)

123. Johnston, "Lennon Sees a Wide Impact in Ouster," 40.

124. Riggs, "In My Ears and in My Eyes (Part 2)."

125. Flynn, *The Truth about Pearl Harbor*.

126. Karnow, *Vietnam*, 380–92.

127. "It is a double-edged paean to dualities, as the title more than suggests." Austin, liner notes, *How Can You Be in Two Places at Once*.

128. Ossman, liner notes, *How Can You Be in Two Places at Once*.

129. The columnist Stewart Alsop added, "In Chicago for the first time in my life it began to seem to me possible that some form of American fascism may really happen here." Perlstein, *Nixonland*, 327, 335.

130. Simpson, *Science of Coercion*, 24.

131. Braddock, "How to Be in Two Places at Once."

132. Williams, "Gravity's Python," 112.

133. Kruse, "The Escalator Ride That Changed America."

CHAPTER 3

1. Caute, *The Year of the Barricades*, 449.

2. Perlstein, *Nixonland*, 472–74.

3. Roger Steffens, quoted at the end of chapter 1, claims to have made hundreds of Firesign Theatre tapes for soldiers in the field, as well as recordings of other music, during his two years in Saigon. Braddock, "How to Be in Two Places at Once."

4. Harmetz, *The Making of "The Wizard of Oz,"* 303–4.

5. Sontag, "Happenings."

6. Bazin, "On the *politique des auteurs*," 251.

7. Bergman explicitly names Burroughs, and Proctor and Ossman later riff on "The Invisible Generation" section of *The Ticket That Exploded* ("get it out of your head and into the machines") in a group interview with Michael Canterbury (May 24, 1970). Burroughs exhorts readers to "turn off the sound track on your television set and substitute an arbitrary sound track prerecorded on your tape recorder street sounds music conversation recordings of other television programs you will find that the arbitrary sound track seems to be appropriate." Firesign had used this exact technique on their *Radio Hour Hour* program on KPPC. Firesign Theatre, *Duke of Madness Motors*, DVD; Burroughs, 213, 205.

8. Ward, "Through Tirebiter's Television," 32, 34; Beard, "The Firesign Theatre."

9. Christgau, "Consumer Guide," Nov. 19, 1970.

10. Ward, "Through Tirebiter's Television," 34, 32.

11. Marcus, "The Firesign Theatre," 130.

12. Ingham, "Art Is Cheese Made Visible," 39. Still another reading connects it to Bob Dylan's *Blonde on Blonde* (1966), whose inside sleeve features Dylan holding an old photograph of a diminutive woman while also clutching a pair of pliers.

13. Williams, "Impressions of U.S. Television," in *Raymond Williams on Television*, 25.

14. The album's working title was *We'll Be Hieronymus Bosch in Jest a Minute, but Faust.*

15. Williams defines flow as "the replacement of a programme series of timed sequential units by a flow series of differently related units in which the timing, though real, is undeclared, and in which the real organisation is something other than the declared organisation." Austin writes, "Commercial television is constantly punctuated by interruptions of one sort or another and . . . as much or more truth is contained in the interruptions as in the so-called content." Williams, *Television*, 93; Austin, liner notes, *Don't Crush That Dwarf.*

16. "A Life in the Day" was broadcast on KRLA live from the Magic Mushroom on January 14, 1968. Firesign debuted "The TV Set" at the Ash Grove in 1969 and performed it throughout their East Coast tour in spring 1970. Firesign Theatre, *Exorcism in Your Daily Life*, 174–97. A recording of "A Life in the Day" is on *Live at the Magic Mushroom*; the Columbia University recording of "The TV Set" is on *Before They Changed the Water.*

17. Williams, "Impressions of U.S. Television," *Raymond Williams on Television*, 24.

18. Spigel, *TV by Design*, 227. *Variety* quoted in Christensen, *America's Corporate Art*, 245.

19. Ossman, interview by the author, April 14, 2016.

20. Bergman would remember, "We were the only people writing on the entire MGM lot, and we used to wander that place. Except for a few security guards, there was nobody there. We wrote in the Thalberg Building, and we were completely alone. We used to have lunch down in the old Copenhagen harbor, or over in the old *National Velvet* paddock, or go down the *Meet Me in St. Louis* street." Wiebel, *Backwards into the Future*, 73–74.

21. Bingen, Sylvester, and Troyan, *MGM*, 270–79; Monaco, *History of the American Cinema*, 38–39. On Hollywood's 1960s' conglomeration see Connor, *Hollywood Math and Aftermath*, 55–56.

22. Firesign Theatre, *Radio Hour Hour*, April 26, 1970, *Duke of Madness Motors*, DVD.

23. Verma, *Theater of the Mind*, 19–21.

24. James, *Rock 'n' Film*, 225.

25. Ossman, "Zack: A Look Back," 3. The rewriting project came to the Firesign Theatre by way of their contacts with Bergman's Amsterdam friends Marijke Koger and Simon Posthuma. Now working as the design collective the Fool, they had designed the clothes the Beatles wore for the satellite broadcast of "All You Need Is Love" (July 1967) and film of "I Am the Walrus" (September 1967), and painted the facade of the Apple Boutique (December 1967).

26. Greenspun, "Screen: 'Zachariah,' an Odd Western," 20.

27. Kael's broadly critical description of *El Topo* might also describe the Firesign Theatre's Columbia albums: "the film is commercialized Surrealism. . . . I think . . . that the counter-culture has begun to look for the equivalent of a drug trip in its theatrical experiences." Kael, "El Poto-Head Comics," 215, 220. Peckinpah's edit of *Pat Garrett and Billy the Kid* was finally released to critical acclaim in 1988. Seydor, *The Authentic Death and Contentious Afterlife of Pat Garrett and Billy the Kid*.

28. Hence the subtitle of Biskind's *Easy Riders, Raging Bulls: How the Sex-Drugs-and-Rock-'n'-Roll Generation Saved Hollywood*.

29. "Another thing that interests me about the Eagles is that I hate them." Christgau, "Trying to Understand the Eagles."

30. While it is true that the name of the band, the James Gang, invoked the West, its iconology was forged in the same spirit that defined the acid-western: the back cover of their 1970 album *Rides Again* directly quoted *Easy Rider*, and the cover of their 1971 live album shows horses tied up at the front of a concert hall (on the reverse, the band is seen scooping up horse shit). Moreover, the James Gang was from Cleveland—one of rock and roll's sacred cities and far from Hollywood.

Walsh, a graduate of Kent State University, did not join the Eagles until 1975, first appearing on the *Hotel California* album (1976). The *Rides Again* iconography was also riffing on the poster insert that accompanied the Jimi Hendrix Experience's *Smash Hits* LP (released in 1968, a year before *Easy Rider*), which showed the band dressed in western gear and riding horses on an old film set.

31. "Pop Records," 60.

32. James's impressive *Rock 'n' Film* is historically delimited so as to avoid rock operas, though there is one substantial note (449n1).

33. Willis, "Rock, Etc.," July 12, 1969, 64.

34. See Frith's discussion, in which he names Pete Townshend British pop's "smartest theorist." "Rock and the Politics of Memory," 63.

35. Christgau, *Any Old Way You Choose It*, 65.

36. "More acts are 'discovered' or 'created' in L.A. and more records are cut in L.A. than in almost all other cities in the world combined." But the piece also made ominous reference to the new category of the "company freak" or "house hippie," "the record company's equivalent of the 'necessary Negro.'" Hopkins, "Los Angeles Scene," 11.

37. Connor names 1970 as "a rough beginning to [Hollywood's] neoclassical era that . . . largely consolidated by 1975." By that time, notably, "Hollywood was about to turn over its prestige productions to genres that had been B-picture staples. . . . It was entering an era where it would regularly risk schlock in its drive for retro appeal." Connor, *The Studios after the Studios*, 17, 57.

38. In addition to Zappa's famous domicile, Cass Elliott lived in Natalie Wood's former home, Denny Doherty had bought Mary Astor's old house, Eric Burdon lived in the Boris Karloff mansion, and John and Michelle Phillips lived in MGM star Jeanette MacDonald's house in Bel Air. Hoskyns, *Waiting for the Sun*, 181.

39. Hopkins, "Los Angeles Scene," 11.

40. "Pop Records," 61. Notable stand-up records of the period include Lily Tomlin, *And That's the Truth* (1972); George Carlin, *FM&AM* (1972), *Class Clown* (1972), and *Occupation: Foole* (1973); and Richard Pryor, *That N_____'s Crazy* (1974). In 1971, Christgau had written, "I strongly suspect the new mainstream of draining back toward an individualism that rock and roll once seemed to challenge." *Any Old Way You Choose It*, 232.

41. Eisenberg, *The Recording Angel*, 94.

42. Bazin, "On the *politique des auteurs*," 251. Subsequent citations of pagination from this source are given parenthetically in the text.

43. Emerick and Massey, *Here, There and Everywhere*, 111–13, 137–41, 241–42.

44. Quoted in Wald, *How the Beatles Destroyed Rock 'n' Roll*, 242.

45. On the transforming role of the sound engineer in studio recording between 1965 and 1975, see Horning, *Chasing Sound*, 104–20, 171–87; and Kealy, "From Craft to Art."

46. Sterne, *The Audible Past*, 182–83.

47. Gunning, "The Cinema of Attraction."

48. Horning, *Chasing Sound*, 121–24; McClay, "Industry's All-Stereo Push Puts the Needle in Consumer Instead of Inbetween the Grooves."

49. In the words of *Abbey Road* engineer Ken Scott,

> As people in England were still not into stereo up to this point . . . , the Beatles had only been in the studio for the mono mixes. . . . But for the White [A]lbum the stereo mixes were often done immediately after the mono version with differences requested by the band members. I was told by Paul that the reason for this was that they had been receiving fan mail telling them how much the fans liked the differences and so they thought they might sell even more records by purposely changing things between mixes.
>
> (Ken Scott, interview by Jason Barnard, Sept. 20, 2018)

50. Altman and Handzo, "The Sound of Sound," 70.

51. Ondaatje and Murch, *The Conversations*, 166; Sergi, *The Dolby Era*, 17.

52. Dolby, "Audio Noise Reduction," 30.

53. Holzman and Daws, *Follow the Music*, 216.

54. Eisenberg, *The Recording Angel*, 89. From Ray's point, it is impossible to resist making the final step to point out the obvious affinities between Derrida's concept of phallogocentrism and the authenticity culture of cock rock. Ray, "Tracking," 142.

55. Sergi, *The Dolby Era*, 92–99. See also Beck, *Designing Sound*, 87–109, 153–202.

56. Colker, "Ray Dolby Dies at 80." See also Sergi, *The Dolby Era*, 19.

57. Doane, "The Voice in the Cinema," 35, 36.

58. David Ossman, personal correspondence, Jan. 30, 2017.

59. Marcus, introduction to Firesign Theatre, *Marching to Shibboleth*, 6.

60. Ossman, interview by Kurt Ericson, May 7, 2015. See also Kealy's discussion of the engineer's transforming role during this period in "From Craft to Art," 213.

61. See Verma, *Theater of the Mind*, 42, 140–41.

62. It was Murch who suggested to George Lucas that the radio, heard in cars and windows throughout *American Graffiti*, should form the film's "backbone." Ondaatje and Murch, *The Conversations*, 119.

63. McLuhan, *Understanding Media*, 23.

64. Quoted in Ingham, "Art Is Cheese Made Visible," 8.

65. Firesign Theatre, *Radio Hour Hour*, May 3, 1970, *Duke of Madness Motors*, DVD.

66. In May 2017, when I was listening to the record with David Ossman, he exclaimed at this point: "Thank God for the modulation—you know you're going someplace, finally!" Ossman, interview by the author, May 19, 2017.

67. *Don't Crush That Dwarf* working script, box 7, folder 8, The Firesign Theatre Collection, National Audio-Visual Conservation Center, Library of Congress.

68. The Electrician is a figure in Burroughs's *Nova Express*, but Firesign may also have been borrowing this extensive electrical power/social power pun from Ralph Ellison's *Invisible Man* (1951), where it is one of the dominant figures in the book. Ford, "Crossroads and Cross-Currents in *Invisible Man*."

69. Lott, *Black Mirror*, 139–56. Here it is also important to remark that there is one instance of a full blackface performance in the Firesign corpus, which is Peter Bergman's performance in the lip-synched film of the 1974 album *Everything You Know Is Wrong*. Bergman appears as "Uncle Tom," returning to inform the white plantation owners who believe they have won the Civil War that they will all now "take turns" at being enslaved, in the album's satiric New Age counterhistorical propaganda film. Providing further context for this performance, though certainly not excusing it, is Bergman's admission in a 2004 interview that his grandfather, a Jewish immigrant named Abe Bergman, had been an end man in a late nineteenth-century minstrel show in New York City, where, according to Bergman's father, he performed with Sitting Bull. Bergman, interview by Maryedith Burrell, Feb. 1, 2004.

70. Firesign Theatre, *Radio Hour Hour*, May 17, 1970, and May 24, 1970, *Duke of Madness Motors*, DVD.

71. The Firesign Theatre discussed *The Hour of Power* on the May 17 *Radio Hour Hour* broadcast. Hartig, "'A Most Advantageous Spot on the Map,'" 307–8; Artman, *The Miracle Lady*.

72. Kruse, *One Nation under God*, 242–47, 260–63.

73. Thompson, "Memo from the Sports Desk," 24.

74. This gag appears to have originated in the March 22 *Radio Hour Hour* when Phil Austin noted that just as God is the word *dog* backwards, Jesus is *sausage* backwards.

75. Willis, "Into the Seventies, for Real," 169.

76. At nearly the same time, Pink Floyd were creating a similar diegetic effect on the *Atom Heart Mother* track "Alan's Psychedelic Breakfast." Later, on *Wish You Were Here* [1975], they would reproduce the effect of a radio playing the title song cross-faded into the "real space" of the song's performance, an inversion of the Tirebiter effect described above. See Mason, *Inside Out*, 135.

77. Austin, liner notes for *Don't Crush That Dwarf*.

78. Hill, "Chanting the Square Deific," 39.

79. Ossman, interview, May 19, 2017.

80. "The Firesign Theatre," *Columbia Daily Spectator*, 39; Wiebel, *Backwards into the Future*, 110.

81. Pansey, "A Vest Has No Sleeves," 45–46

82. Antin, *Code of Flag Behavior*. Remarkably, the archival material recited by Firesign is drawn from several sources. One source is Tom B. Underwood

et al., *Cherokee Legends and the Trail of Tears*, 21–22. The Free Music Store performance was broadcast on the March 8 *Radio Hour Hour* (*Duke of Madness Motors*, DVD).

83. On February 9, an all-women's issue of the underground paper *Rat* had hit the streets of New York. Gitlin, *The Sixties*, 373.

84. Carson, *In Struggle*, 297.

85. KPPC rebroadcast the March 10 episode of Paul Gorman's WBAI program five days later on Firesign's *Radio Hour Hour*. "New New Ork," *Duke of Madness Motors*, DVD.

86. *Fusion* magazine interview, Sept. 1970, *Duke of Madness Motors*, DVD.

87. Rudd, *Underground*, 189.

88. Thanks to Claudia Verhoeven for making this connection for me. Deutsch, "Bizarre Humor at Grim Tate Trial," 7; Associated Press, "Laughter in a Strange Setting," 11.

89. Hill, "Chanting the Square Deific," 40.

90. Gendron, *Between Montmartre and the Mudd Club*, 29, 73.

91. On the surrealists' sustained interest in cinema, see Hammond, *The Shadow and Its Shadow*; and Kyrou, *Le Surréalisme au Cinema*.

92. Sontag, "Happenings," 269.

93. Cushing, "Outline of Zuñi Creation Myths," quoted in Wright, *Clowns of the Hopi*, 69.

94. Sontag, "Happenings," 273 (my emphasis).

95. Casale, quoted in Reynolds, *Rip It Up and Start Again*, 76–77.

96. Buckley, "H Bomb."

CHAPTER 4

1. Austin, liner notes, *Don't Crush That Dwarf.*

2. Williams, *Television*, 19.

3. Williams, 20–21.

4. Hoeffler, "Silicon Valley U.S.A."; Eco, "Travels in Hyperreality," 47.

5. Ossman, interview by the author, April 14, 2016.

6. Marsh, rev. of *I Think We're All Bozos on This Bus*, 60.

7. Tris, "Firesign Theatre at the Ash Grove," 54. This section of the chapter takes its heading from Willis, "Rock Etc.: Into the Seventies, for Real."

8. Tiegel, "School of Relevant Humor Makes the Campus a Fun Place to Study," 31.

9. See Jessen, "An Avant-Garde Religious Radio Happening"; and the Live Earl Jive's short essay in *Duke of Madness Motors* (booklet), 48–50.

10. Proctor and Schreiber, *Where's My Fortune Cookie?*, 100.

11. Fong-Torres, "Wacco's Circle Is Unbroken," 18. According to *Creem*, the *Big Suitcase* was to have been "a pastiche of effects; all in 35mm., naturally, but part will be color, part black and white, with a portion of the color done with an old Technicolor tri-pack camera, to achieve the same Technicolor effect that is displayed in the early fifties' films. . . . The sound-track will be done in 8-track stereo." "Firesign Filming," 13.

12. Simon, *Truth, Lies & Hearsay*, 165, 181.

13. Hepworth, *Never a Dull Moment*, 99.

14. "Top LP's," Oct. 30, 1971, 66; Hill, "Chanting the Square Deific," 38–40.

15. Christgau also wrote, "This is everything you would expect from the Firesign Theatre but funny, which is something like saying the Stones had a great session only Bill and Charlie stayed home." Christgau, "Consumer Guide," Dec. 12, 1971, and n.d. 1972; Greil Marcus, "Firesign Theatre," 130.

16. Neill, rev. of *I Think We're All Bozos on This Bus*, 18.

17. Young, rev. of *Dear Friends*, 50; Gold, "Muzak of the Spheres," Oct. 18, 1972, 21.

18. Marsh, rev. of *I Think We're All Bozos on This Bus*, 60.

19. Land, *Active Radio*, 117–19; see also Gitlin, *The Whole World Is Watching*.

20. Bodroghkozy, *Groove Tube*, 98–100.

21. Willis, "Herbert Marcuse, 1898–1979," 144.

22. Gitlin, *The Sixties*, 417.

23. The *San Francisco Chronicle* called the Bed-Ins "A Put-On Honeymoon." Fong-Torres, *The Rolling Stone Rock 'n' Roll Reader*, 30. Lennon's comments can be seen in Steve Gebhardt's *Ten for Two*.

24. Thompson, "Memoirs of a Wretched Weekend in Washington," 205.

25. Hochman, *The Listeners*, 173–219.

26. Gitlin, *The Sixties*, 413.

27. Kauffman, *Direct Action*, 37.

28. Ossman, interview, April 14, 2016.

29. David Ossman and Philip Proctor, personal correspondence, Sept. 13, 2017.

30. Ritter, "Hamburger All Over the Info.Highway," 8. Ritter also points out that then-current Unix computers had access to a program ("ching") that was the full text of Richard Wilhelm's translation of the *I Ching*: "the Unix Manual pages for ching actually quotes [the lines quoted by Nick Danger on *How Can You Be*], which makes us suspect that FT fans go deep into the legions of hackers responsible for the development of Unix."

31. Ossman, interview, April 14, 2016.

32. The album's title punned on the "bus," studio terminology for a channel that runs into or out of a stereo mixing console, and it also referenced the

psychedelic bus "Furthur," in which Ken Kesey and the Merry Pranksters traversed the country in the mid-1960s. Finally, it was inspired by the new transportation methods (the "intramural bus") at the 1933 Chicago World's Fair.

33. Returning to record the Beatles' final album in 1969, Geoff Emerick describes in very similar terms the EMI studio's new transistor-based recording console: "I preferred the punchier sound we had gotten out of the old tube console and four-track recorder; everything was sounding mellower now. . . . softer and rounder. It's subtle, but I'm convinced that the sound of the new console and tape machine overtly influenced the performance and the music on *Abbey Road.*" Emerick and Massey, *Here, There and Everywhere*, 277–78. See also Milner, *Perfecting Sound Forever*, 160; and Womack, *Solid State*.

34. Marcus, *Mystery Train*, 71–91.

35. Roszak, *The Making of a Counterculture*, 6.

36. Hughes, *Chasing Shadows*, 2; Kelley, "Ron Ziegler, Press Secretary to Richard Nixon, Is Dead at 63."

37. Greenhalgh, *Fair World*, 15.

38. Greenhalgh, 83, 147. Firesign appears not to have known that the Chicago fair also featured the first demonstration of binaural audio (on a Bell Laboratories robot with microphone ears). Théberge, Devine, and Everrett, introduction to *Living Stereo*, 17.

39. Ganz, *The 1933 Chicago World's Fair*, 2–3; *Official Guide Book of the Fair 1933*; *Chicago: A Century of Progress, 1833–1933*.

40. Zuboff, *The Age of Surveillance Capitalism*, 15.

41. *Chicago: A Century of Progress*, 41.

42. Tozer, "A Century of Progress, 1833–1933."

43. Kurlansky, *1968*, 194; Kauffman, *Direct Action*, 15–16.

44. *Chicago: A Century of Progress*, 44–45.

45. Proctor, interview on *Comedy on Vinyl*, episode 269.

46. Eco, "Travels in Hyperreality," 3, 45.

47. Bemis, *Disney Theme Parks and America's National Narratives*, 42.

48. Tarnoff, "Weizenbaum's Nightmares."

49. Wardrip-Fruin, *Expressive Processing*, 30.

50. Weizenbaum, "ELIZA," 36.

51. Weizenbaum, 36.

52. Weizenbaum, "Contextual Understanding by Computers," 475.

53. Weizenbaum, "ELIZA," 37.

54. Grobe, "The Programming Era."

55. Weizenbaum, "Contextual Understanding by Computers," 475.

56. Quoted in Perlstein, *Nixonland*, 333.

57. David Ossman and Philip Proctor, personal correspondence, Sept. 13, 2017.

58. Lewisohn, *The Beatles Recording Sessions*, 72.

59. Combined with other data, these profiles now allow tech industries and governments "to be able to fix the identities of their speakers, track them, connect their vocalizations with other data, and thereby build a bigger, more sophisticated profile of the whole person, including the person's emotional states, desires, and intentions." Sterne and Sawheny, "The Acousmatic Question and the Will to Datafy," 292.

60. Roose, "Bing's A.I. Chat."

61. Ginsberg, "Wichita Vortex Sutra," *Planet News*, 127.

62. Weizenbaum, *Computer Power and Human Reason*, 201.

63. Weizenbaum, "ELIZA," 43.

64. Weizenbaum, *Computer Power and Human Reason*, 5.

65. Quoted in Tarnoff, "Weizenbaum's Nightmares."

66. Weizenbaum, *Computer Power and Human Reason*, 5–8. Subsequent citations of pagination from this source are given parenthetically in the text.

67. Many of these ideas would be taken up more systematically by Paul N. Edwards in *The Closed World*.

68. Hill, "Chanting the Square Deific," 39.

69. Goodyear, "From Technophilia to Technophobia," 170; Tuchman, *A Report on the Art and Technology Program of the Los Angeles County Museum of Art*.

70. Fallon, *Creating the Future*, 9–25.

71. Kozloff, "The Multimillion Dollar Art Boondoggle," 76.

72. Heibel, "LACMA Launches Art + Technology Lab."

73. Malone, "Postmark: Deep Space," 7.

74. Proctor and Schreiber, *Where's My Fortune Cookie?*, 120.

75. Simon, "Who Still Needs the Carnivalesque?"

76. Bender et al., "Stochastic Parrots"; Phan, "Amazon Echo and the Aesthetics of Whiteness."

77. Crawford and Joler, "Anatomy of an AI System."

78. Tarnoff, "Weizenbaum's Nightmares."

CHAPTER 5

1. Shamberg, *Guerrilla Television*, 74.

2. Smith, *Spoken Word*, 178–79; Fiske, *Television Culture*, 104–5.

3. Uricchio, "Television's Next Generation," 181n10.

4. Uricchio, 165.

5. Brundson, "What Is the 'Television' of Television Studies?"

6. They made one short film, "The Bob Sideburn News," for an LA television show in 1971. Working without Ossman in the 1980s, the Firesign Theatre produced works for home video: *Nick Danger and the Case of the Missing Yolk*

(1983) and *Eat or Be Eaten*, which accompanied their first album and was first broadcast on Cinemax in 1985. There were also two video projects that dubbed new soundtracks over collages of old movies: Proctor and Bergman's *J-Men Forever* (1979) and the three-piece Firesigns' *Hot Shorts* (1985).

7. Braddock and Morton, "How to Make Hyperobject Sound Art," 54.

8. Bodroghkozy, *Groove Tube*, 98–106.

9. Spigel, *TV by Design*, 188–89.

10. Ward, "Why Do Kids Love These Four Zany Guys," D33; Garson, "Luddites in Lordstown," 71–72.

11. Bronson, "The New Comedy," 10–11.

12. Ossman, "The Martian Space Party Diary."

13. Devine, "Decomposed," 375–76.

14. Walker, *What You Want Is in the Limo*, 98–99.

15. Walker, 115–16; Povey, *Echoes*, 154.

16. "Firesign World / Anytown USA," n.p.

17. Andrew, *Anyway, Anyhow, Anywhere*, 251.

18. Ossman, *Dr. Firesign's Follies*, 47, 49.

19. Wiebel, *Backwards into the Future*, 236.

20. Willis, "Rock, Etc.," August 12, 1972, 57; Willis, "Creedence Clearwater Revival," 326.

21. "Random Notes," 4.

22. The complete broadcast can be heard on the *Duke of Madness Motors* DVD.

23. Parks, *Cultures in Orbit*, 21–46.

24. Kurlansky, *1968*, 40, 52. See also Hallin, *The "Uncensored War."*

25. Barnouw, *Tube of Plenty*, 49–50.

26. Firesign Theatre Collection, National Audio-Visual Conservation Center, Library of Congress, box 106.

27. Williams, *Television*, 149.

28. Parks, *Cultures in Orbit*, 174.

29. *Son of Godzilla*, notably, was released straight to television in the US. Kalat, *A Critical History and Filmography of Toho's Godzilla Series*, 98–110.

30. This phrase comes from Eric Lott's foundational study of blackface minstrelsy, *Love and Theft*. Jack Hamilton's description of the Rolling Stones' relationship to Black American blues music could also describe Firesign's use of Japanese media culture in 1971: "a chain reaction marked by dialectical flickerings of expropriation and homage, fetishization and appreciation, opportunism and guilt." Hamilton, *Just Around Midnight*, 119.

31. James, "The Hard Way to Make Money in Television," 44.

32. Wald, *It's Been Beautiful*.

33. See Richard Dyer's classic critique in *White*, 1–40.

34. Quoted in Merline, "The Firesign Theatre," n.p.

35. Parks, *Cultures in Orbit*, 88–98, 160–64.

36. Ossman, *Dr. Firesign's Follies*, 49.

37. Ossman, 50.

38. Steyerl, "The Spam of the Earth," 161.

39. "Rock-a-Rama," 75; Dupree, rev. of *Not Insane*, 63.

40. Herbert, *Maverick Movies*, 30.

41. Williams, *Raymond Williams on Television*, xi.

42. Lotz, *We Now Disrupt This Broadcast*, 190–91.

43. Mullen, *The Rise of Cable Programming in the United States*, 64.

44. McMurria, *Republic on the Wire*, 32–33.

45. Hoffman, *Revolution for the Hell of It*, 168.

46. Sloan Commission, *On the Cable*, 167–68.

47. Mullen, *The Rise of Cable Programming in the United States*, 6.

48. Streeter, *Selling the Air*, 175–76.

49. Streeter, "Blue Skies and Strange Bedfellows," 238–40.

50. Williams, *Raymond Williams on Television*, xi.

51. Whereas Williams remarks on its possibilities for education, it was met with alarm in the *Los Angeles Free Press*, which foresaw the way "it may lead to fantastically exploitative set-ups for shopping and marketing." Williams, *Television*, 144; *LA Free Press*, quoted in Bodroghkozy, *Groove Tube*, 55–56.

52. Williams, *Television*, 144–47.

53. Uricchio, "Television's Next Generation," 167.

54. Ossman, interview by the author, April 14, 2016.

55. Callwood, "My Favorite Album of All Time."

56. Williams, *Television*, 91–93.

57. Adorno, "How to Look at Television," 220; Adorno, "Prologue to Television," 49.

58. Villarejo, *Ethereal Queer*, 30–65.

59. Firesign Theatre Collection, National Audio-Visual Conservation Center, Library of Congress, box 13, folders 1–2.

60. Shoe, "On the Campoon Trail," 9.

61. Williams, "Drama in a Dramatised Society," in *Raymond Williams on Television*, 4.

62. Villarejo, *Ethereal Queer*, 107.

63. Mead, "As Significant as the Invention of Drama or the Novel."

64. Ebert, rev. of *Billy Jack*.

65. Hodenfield, "What Price Ecstasy?"

66. The Cherokee Apache activist Frank Clearwater (named fictionally as the winner of the Academy Award for *I Gave it Back*) had been shot and killed by U.S. Marshals in April 1973.

67. Tilly, "Collective Violence in European Perspective," 9.

68. McCarthy, *Ambient Television*.

69. Cavell, "The Fact of Television," 75. Subsequent citations of pagination from this source are given parenthetically in the text.

CODA

1. Haggins and Lotz, "At Home on the Cutting Edge," 151–71.

2. See also the same year's *Good-bye Pop 1952–1976* by *National Lampoon*.

3. In Jacob Smith's excellent discussion, Martin's catchphrase-based comedy addressed a mass audience rather than the "counterpublics" that had been recorded comedy's audience before and during the Firesign era. Smith, *Spoken Word*, 196.

4. Robert Plant, interview, Feb. 20, 1970, 11.

5. At the end of the decade, Howard Hesseman, who had been a founding member of the improv group The Committee and a DJ on San Francisco's legendary freeform station KMPX, ended up playing a parody of himself on the CBS sitcom *WKRP in Cincinnati*.

6. Uhelszki, "In the Beginning," 6.

7. Bangs, "Not Insane. And Not Funny Either," 20.

8. Walls, rev. of *In The Next World, You're on Your Own*, 68.

9. Marsh and Stein, *The Book of Rock Lists*, 285.

10. Weisbard, *Top 40 Democracy*.

11. Firesign's video-game work dates to the aborted 1981 project *Pink Hotel Burns Down*. It was later realized as the 1985 album *Eat or Be Eaten*, whose CD release included subcode graphics and was accompanied by an eponymous broadcast on Cinemax. A decade later, Peter Bergman would write and release *Pyst*, a computer game that was a critical parody of the popular adventure game *Myst*.

12. Hiss and McClelland, "The Firesign Comics Are Hot Stuff," D5–6.

13. Karpel, "George Who for Veep," 49.

14. Edgar Bullington to Tom Gedwillo, Dec. 5, 1974, box 62, folder 1, Firesign Theatre Collection, National Audio-Visual Conservation Center, Library of Congress.

15. Jenkins, *Textual Poachers*, 53.

16. Radway, "Zines, Half-Lives, and Afterlives," 140–50.

17. Conliff, "Everybody Needs Nobody Sometimes," 25.

18. Riggs, "Postmark: Deep Space," 4.

19. According to Firesign's *Big Book of Plays*, "the Bozos have learned to enjoy their free time, which is all the time." Firesign Theatre, *Marching to Shibboleth*, 101. On the Grateful Dead's 1972 European tour, the bus reserved for the band

(as opposed to the one for the roadies) was the "Bozo" bus. Hickey, "'Dark Star' by the Grateful Dead."

20. Cusack, *Invented Religions*, 83–110.

21. When I cold-called Negativland's Don Joyce in the 1990s, he was evasive on the group's relationship to the Firesign Theatre, understandably stressing the differences in their work. Some years later, he embraced the connection much more openly: "I'm flattered to pieces really since they have been an inspiration to me since I began buying their records in the 60s, and I would credit those records with being a primary inspiration for my getting into radio and doing funny production stuff there, as well as impressing me with how OFF THE WALL INTELLEC-TUAL entertainment could actually be." Don Joyce, "Negativland Interview."

22. Rose, *Black Noise*, 73.

23. Rose, 90.

24. Perchard, "Hip Hop Samples Jazz," 300.

25. Quoted in Perchard, 290.

26. On the SF turntablist scene, see Katz, *Groove Music*, 127–52, 179–213.

27. Richard Metzger, personal correspondence, August 31, 2021.

28. Markle, "The Hip-Hop DJ as Black Archaeologist," 208; Rose, *Black Noise*, 79.

29. Perchard, "Hip Hop Samples Jazz," 290.

30. Tate, "Yabba Dabba Doo-Wop," in *Flyboy in the Buttermilk*, 137–41; Sean O'Hagan, "De La Soul."

31. Dery, "Digital Underground, Coldcut and De La Soul Jam the Beat."

32. Rose, *Black Noise*, 85, 95.

33. Rubin, Michaels, Merline, Rice, and Henssler, "Rock Lit"; *Nuts to You.*

Bibliography

"Actor Peter Fonda Is Taken into Custody Near the Fifth Estate." *Los Angeles Free Press*, Dec. 2, 1966.

Adorno, Theodor W. "How to Look at Television." *Quarterly of Film, Radio, and Television* 8, no. 3 (1954): 213–35.

———. "Prologue to Television." In *Critical Models: Interventions and Catchwords*, translated by Henry W. Pickford, 49–58. New York: Columbia University Press, 1998.

Altman, Rick, and Stephen Handzo. "The Sound of Sound: A Brief History of the Reproduction of Sound in Movie Theaters." *Cinéaste* 21, no. 1-2 (1995): 68–71.

Anderson, Tim J. *Making Easy Listening: Material Culture and Postwar American Recording*. Minneapolis: University of Minnesota Press, 2006.

Andrew, Neil. *Anyway, Anyhow, Anywhere: The Complete Chronicle of the Who, 1958-1978*. London: Virgin, 2002.

Antin, David. *Code of Flag Behavior*. Los Angeles: Black Sparrow, 1968.

Artaud, Antonin. *The Theater and Its Double*. Translated by Mary Caroline Richards. New York: Grove Weidenfeld, 1958.

Artman, Amy Collier. *The Miracle Lady: Kathryn Kuhlman and the Transformation of Charismatic Christianity*. Grand Rapids, MI: William B. Eerdmans, 2019.

Ashbery, John. *Commotion of the Birds*. New York: Ecco, 2016.

Ashby, Arved. *Absolute Music, Mechanical Reproduction*. Berkeley: University of California Press, 2010.

Associated Press. "Laughter in a Strange Setting: The Tate Trial." *Eureka (CA) Times Standard*, Sept. 8, 1970.

The Astrology Album: Your Horoscope and Character Analysis in Music and Narration. Columbia, CL 2689, 1967, LP.

Austin, Phil. Liner notes for Firesign Theatre, *Don't Crush That Dwarf, Hand Me the Pliers*. Mobile Fidelity Sound Lab MFCD 880, 1987. Originally released in 1970 by Columbia Records.

———. Liner notes for Firesign Theatre, *How Can You Be in Two Places at Once When You're Not Anywhere at All*. Mobile Fidelity Sound Lab MFCD 834, 1988. Originally released in 1969 by Columbia Records.

Bakhtin, Mikhail. *The Dialogic Imagination*. Edited by Michael Holquist. Translated by Caryl Emerson and Michael Holquist. Austin: University of Texas Press, 1981.

Bangs, Lester. "Not Insane. And Not Funny Either. The Firesign Theatre's Catalog of Misconceptions." *Creem*, April 1975.

Barbrook, Richard, and Andy Cameron. "The Californian Ideology." In *The Internet Revolution: From Dot-com Capitalism to Cybernetic Communism*, 12–27. Amsterdam: Institute of Network Cultures, 2015.

Barnouw, Erik. *The Golden Web: A History of Broadcasting in the United States*. Vol. 2, *1933 to 1953*. New York: Oxford University Press, 1968.

———. *Tube of Plenty: The Evolution of American Television*. 2nd rev. ed. Oxford: Oxford University Press, 1990.

Barrett, Don. "Lew Irwin Sets the Record Straight on Origins of 1110/KRLA Credibility Gap." *570KLAC* (blog). July 15, 2010. https://krlabeat.sakionline .net/img/lewirwin.pdf.

Barry, Eric D. "High-Fidelity Sound as Spectacle and Sublime, 1950–1961." In *Sound in the Age of Mechanical Reproduction*, edited by David Suisman and Susan Strasser, 115–38. Philadelphia: University of Pennsylvania Press, 2010.

Baudrillard, Jean. *Simulacra and Simulation*. Translated by Sheila Faria Glaser. Ann Arbor: University of Michigan Press, 1994.

Bazin, André. "On the *politique des auteurs*." In *Cahiers du Cinéma: The 1950s—Neo-Realism, Hollywood, New Wave*, edited by Jim Hillier, 248–59; translated by Peter Graham. Cambridge, MA: Harvard University Press, 1985.

Beard, M. C. "The Firesign Theatre: A Review." *College English* 33, no. 3 (Dec. 1971): 379–82.

Beck, Jay. *Designing Sound: Audiovisual Aesthetics in 1970s American Cinema*. New Brunswick, NJ: Rutgers University Press, 2016.

Belz, Carl. *The Story of Rock*. New York: Oxford University Press, 1969.

Bemis, Bethanee. *Disney Theme Parks and America's National Narratives: Mirror, Mirror for Us All.* New York: Routledge, 2023.

Bender, Emily, Angelina McMillan-Major, Timnit Gebru, and Shmargaret Shmitchell. "Stochastic Parrots: Can Large Language Models Be Too Big?" *FAccT '21*, March 3–10, 2021. Virtual Event, Canada.

Berghahn, Volker R. *America and the Intellectual Cold Wars in Europe: Shepard Stone between Philanthropy, Academy, and Diplomacy.* Princeton, NJ: Princeton University Press, 2001.

Bergman, Peter. Interview by Maryedith Burrell, 2003–4. Transcribed by Taylor Jessen. Firesign Theatre Digital Archive, Burbank.

———. "Peter Bergman on the Hopi Indian." Interview by Tom Schulz, Joe Dana, and Teddy Tubbe. *Oracle of Southern California*, March 1967.

Bijsterveld, Karin. "Acoustic Cocooning: How the Car Became a Place to Unwind." *Senses and Society* 5, no. 2 (2010): 189–211.

Biner, Pierre. *The Living Theatre.* New York: Horizon, 1972.

Bingen, Steven, Stephen X. Sylvester, and Michael Troyan. *MGM: Hollywood's Greatest Backlot.* Solana Beach, CA: Santa Monica Press, 2011.

Biskind, Peter. *Easy Riders, Raging Bulls: How the Sex-Drugs-and-Rock-'n'-Roll Generation Saved Hollywood.* New York: Simon & Schuster, 1998.

Black, Austin. *The Tornado in My Mouth.* New York: Exposition, 1966.

Blau, Herbert. *Programming Theater History: The Actor's Workshop of San Francisco.* London: Routledge, 2013.

Blesser, Barry, and Linda-Ruth Salter. *Spaces Speak, Are You Listening? Experiencing Aural Architecture.* Cambridge, MA: MIT Press, 2007.

Bodroghkozy, Aniko. *Groove Tube: Sixties Television and the Youth Rebellion.* Durham, NC: Duke University Press, 2001.

Braddock, Jeremy. "How to Be in Two Places at Once: Listening to the Firesign Theatre in the US and Vietnam." *The Organist.* Podcast. Dec. 14, 2017. https://www.kcrw.com/news-culture/shows/the-organist/be-in-two-places-at-once-the-firesign-theatre.

———. "Marcus/Christgau: Whose Era?" *Journal of Popular Music Studies* 28, no. 1 (March 2016): 126–33.

Braddock, Jeremy, and Timothy Morton. "How to Make Hyperobject Sound Art: Occupying the Electromagnetic Field with the Firesign Theatre." *New Centennial Review* 18, no. 2 (2018): 39–68.

Brodie, Ian. *A Vulgar Art: A New Approach to Stand-Up Comedy.* Jackson: University Press of Mississippi, 2014.

Bromell, Nick. *Tomorrow Never Knows: Rock and Psychedelics in the 1960s.* Chicago: University of Chicago Press, 2000.

Bronson, Hal. "The New Comedy: 1 Part Message to 4 Parts Laughter." *Rock*, April 10, 1972.

Brundson, Charlotte. "What Is the 'Television' of Television Studies?" In *The Television Studies Book*, edited by Christine Geraghty and David Lusted, 95–113. London: Arnold, 1998.

Buckley, Lord [Richard Myrle]. "H Bomb." *Bad Rapping of the Marquis de Sade*. World Pacific Records WPS 21889, 1969, LP.

Buddies, The. "Duckman (Part 1 and 2)." Decca, 31920, 1966, 7".

Burroughs, William S. *The Ticket That Exploded*. New York: Grove, 1967.

Byrds, The. *The Notorious Byrd Brothers*. Columbia, CS 9575, 1968, LP.

———. *Younger Than Yesterday*. Columbia, CL 2642, 1967, LP.

Callwood, Brett. "My Favorite Album of All Time: Blag Dahlia of the Dwarves." *LA Weekly*, July 1, 2019. https://www.laweekly.com/my-favorite-album-blag-dahlia-of-the-dwarves.

Cantril, Hadley, Hazel Gaudet, and Herta Herzog. *The Invasion from Mars: A Study in the Psychology of Panic*. 1940. New Brunswick, NJ: Transaction, 2005.

Carpenter, John. "Dr. Firesign's Travelling Antique Circus." *Los Angeles Free Press*, Oct. 19, 1969.

Carson, Clayborne. *In Struggle: SNCC and the Black Awakening of the 1960s*. Cambridge, MA: Harvard University Press, 1995.

Caute, David. *The Year of the Barricades: A Journey through 1968*. New York: Harper and Row, 1988.

Cavell, Stanley. "The Fact of Television." *Daedalus* 111, no. 4 (1982): 75–96.

Chapple, Steve, and Reebee Garofalo. *Rock 'n' Roll Is Here to Pay: The History and Politics of the Music Industry*. Chicago: Nelson-Hall, 1977.

Chartier, Roger. "Labourers and Voyagers: From the Text to the Reader." In *The Book History Reader*, edited by David Finkelstein and Alistair McCleery, 87–98. 2nd ed. London: Routledge, 2006.

———. "The Practical Impact of Writing." In *The Book History Reader*, edited by David Finkelstein and Alistair McCleery, 118–42. 2nd ed. London: Routledge, 2006.

Cheech and Chong. *Cheech and Chong*. Ode, SP 77010, 1971, LP.

Chicago: A Century of Progress, 1833–1933. Chicago: Marquette, 1933.

Christensen, Jerome. *America's Corporate Art: The Studio Authorship of Hollywood Motion Pictures*. Stanford, CA: Stanford University Press, 2012.

Christgau, Robert. "Album of the Year." *Village Voice*, Jan. 8, 1970. https://www.robertchristgau.com/xg/rock/album-70.php.

———. *Any Old Way You Choose It: Rock and Other Pop Music, 1967–1973*. Baltimore: Penguin, 1973.

———. "Consumer Guide." *Newsday*, n.d., 1972. https://robertchristgau.com/get_artist.php?name=firesign.

———. "Consumer Guide." *Village Voice*, Dec. 11, 1969. https://robertchristgau.com/xg/cg/cg5.php.

———. "Consumer Guide." *Village Voice*, Nov. 19, 1970. https://robertchristgau .com/xg/cg/cg14.php.

———. "Consumer Guide." *Village Voice*, Dec. 12, 1971. https://robertchristgau .com/xg/cg/cg21.php.

———. "Consumer Guide." *Village Voice*, Jan. 17, 1995. https://robertchristgau .com/xg/cg/cgv195-95.php.

———. "Consumer Guide." *Village Voice*, Dec. 26, 2000. https://robertchristgau .com/xg/cg/cgv1200-00.php.

———. "New Kind of Comedy Served on a Platter." *Newsday*, March 3, 1974. https://robertchristgau.com/xg/news/nd740303.php.

———. "Records." *Cheetah*, Oct. 1967.

———. "Trying to Understand the Eagles." *Newsday*, June 1972. https://www .robertchristgau.com/xg/bk-aow/eagles.php.

Christian, T. M. [Greil Marcus and Bruce Miroff]. "The Masked Marauders." Reprint, *Rolling Stone Record Review*, edited by *Rolling Stone* magazine staff, 392–94. New York: Pocket Books, 1971.

Close, Del, and John Brent. *How to Speak Hip*. Mercury OCM 2205, 1961, LP.

Colker, David. "Ray Dolby Dies at 80; Engineer's Sound System Eliminated Underlying Noise." *Los Angeles Times*, Sept. 12, 2013. https://www.latimes .com/local/obituaries/la-me-ray-dolby-20130913-story.html.

Collins, Judy. *Wildflowers*. Elektra, EKS 74012, 1967, LP.

"Comedy and Spoken Word Spotlights—1961." *Billboard Music Week*, Nov. 20, 1961.

Conliff, Steve. "Everybody Needs Nobody Sometimes." *Open Road*, Spring 1977.

Connor, J. D. *Hollywood Math and Aftermath: The Economic Image and the Digital Recession*. New York: Bloomsbury, 2018.

———. *The Studios after the Studios: Neoclassical Hollywood (1970–2010)*. Stanford, CA: Stanford University Press, 2015.

Corwin, Norman. *On a Note of Triumph*. New York: Simon & Schuster, 1945.

———. "We Hold These Truths." In *More by Corwin: 16 Radio Dramas*, 57–87. New York: Henry Holt, 1944.

Crawford, Kate, and Vladan Joler. "Anatomy of an AI System: The Amazon Echo as an Anatomical Map of Human Labor, Data and Planetary Resources." AI Now Institute and Share Lab, 2018. https://anatomyof.ai/.

Credibility Gap, The. *A Great Gift Idea*. Reprise, MS 2154, 1973, LP.

Cull, Nicholas John. *The Cold War and the United States Information Agency: American Propaganda and Public Diplomacy, 1945–1989*. Cambridge: Cambridge University Press, 2008.

Cusack, Carole M. *Invented Religions: Imagination, Fiction and Faith*. Farnham [UK]: Ashgate, 2010.

Cushing, Frank Hamilton, and Frederick Webb Hodge. *Outlines of Zuñi Creation Myths*. Washington, DC: US Government Printing Office, 1896.

Dallas, Paul V. *Dallas in Wonderland: The Pacifica Approach to Free Radio.* Los Angeles: Self-published, 1967.

Davis, Clive, with Anthony DeCurtis. *The Soundtrack of My Life.* New York: Simon & Schuster, 2012.

Davis, Mike. "Riot Nights on Sunset Strip." In *In Praise of Barbarians: Essays against Empire*, 312–29. Chicago: Haymarket, 2007.

Deloria, Philip J. *Playing Indian.* New Haven, CT: Yale University Press, 1998.

Dery, Mark. "Digital Underground, Coldcut and De La Soul Jam the Beat." *Keyboard*, March 1991. *Rock's Backpages.* http://www.rocksbackpages.com /Library/Article/digital-underground-coldcut-and-de-la-soul-jam-the-beat.

Deutsch, Linda. "Bizarre Humor at Grim Tate Trial." *Ontario Daily Report*, Sept. 7, 1970. https://newspaperarchive.com/ontario-daily-report-sep-07-1970-p-7.

Devine, Kyle. "Decomposed: A Political Ecology of Music." *Popular Music* 34, no. 3 (2015): 367–89.

Dick, Philip K. *The Man in the High Castle.* New York: Putnam, 1962.

"Dit Nummer is een Happening." *Ratio* 2, no.1 (1965).

Doane, Mary Ann. "The Voice in the Cinema: The Articulation of Body and Space." *Yale French Studies*, no. 60 (1980): 33–50.

Dolby, Ray M. "Audio Noise Reduction: Some Practical Aspects." *Audio*, July 1968.

Douglas, Susan J. *Listening In: Radio and the American Imagination, from Amos 'n' Andy and Edward R. Murrow to Wolfman Jack and Howard Stern.* New York: Random House, 1999.

Ducksworth-Ford, Robert A. "Column: Is Paul Dead? or Are You?" *Los Angeles Free Press*, Nov. 7–13, 1969.

Duncan, Robert. Review of . . . *Is It Something I Said?*, by Richard Pryor. *Creem*, Nov. 1975.

Dupree, Tom. Review of *Not Insane or Anything You Want To*, by Firesign Theatre. *Creem*, March 1972.

Dyer, Richard. *White.* 20th anniv. ed. London: Routledge, 2017.

Ebert, Roger. Review of *Billy Jack*, dir. T. C. Frank. August 2, 1971. https:// www.rogerebert.com/reviews/billy-jack-1971.

Eco, Umberto. "Travels in Hyperreality." In *Travels in Hyperreality: Essays*, translated by William Weaver, 1–58. San Diego: Harcourt Brace Jovanovich, 1986.

Edmonds, Ben. "The Velvet Underground." *Fusion*, Feb. 5, 1971.

Edwards, Paul N. *The Closed World: Computers and the Politics of Discourse in Cold War America.* Cambridge, MA: MIT Press, 1996.

Eisenberg, Evan. *The Recording Angel: Music, Records, and Culture from Aristotle to Zappa.* New Haven, CT: Yale University Press, 2005.

Elsaesser, Thomas. "Media Archaeology as Symptom." *New Review of Film and Television Studies* 14, no. 2 (2016): 181–215.

Emerick, Geoff, and Howard Massey. *Here, There, and Everywhere: My Life Recording the Music of the Beatles*. New York: Gotham, 2006.

Eno, Brian. "The Studio as Compositional Tool." In *Audio Culture: Readings in Modern Music*, edited by Christopher Cox and Daniel Warner, 185–88. Rev. ed. New York: Bloomsbury, 2017.

Ernst, Wolfgang. "Media Archaeology: Method and Machine vs. History and Narrative of Media." In *Media Archaeology: Approaches, Applications, and Implications*, edited by Erkki Huhtamo and Jussi Parikka, 239–55. Berkeley: University of California Press, 2011.

Esslin, Martin. *The Theatre of the Absurd*. 3rd ed. New York: Vintage, 2004.

Everett, Todd. "Firesign Theatre." *Variety*, Nov. 22, 1993. https://variety.com /1993/legit/reviews/firesign-theatre-1200434234.

Fallon, Michael. *Creating the Future: Art and Los Angeles in the 1970s*. Berkeley, CA: Counterpoint, 2014.

Finkelstein, Sidney. *Sense and Nonsense of McLuhan*. New York: International Publishers, 1968.

"Firesign Filming: 'Funniest Ever.'" *Creem*, March 1971.

"The Firesign Theatre." *Columbia Daily Spectator*, March 6, 1970.

Firesign Theatre. *Duke of Madness Motors: The Complete "Dear Friends" Radio Era, 1970–1972*. Canada: Seeland, 2010. DVD and booklet.

———. *Exorcism in Your Daily Life: The Psychedelic Firesign Theatre at the Magic Mushroom, 1967*. Albany, GA: BearManor Media, 2012.

———. *Marching to Shibboleth: The Big Big Book of Plays*. Introduction by Greil Marcus. Burbank, CA: Firesign Theatre Books, 2013.

———. *Profiles in Barbecue Sauce: The Psychedelic Firesign Theatre on Stage— 1967–1972*. Albany, GA: BearManor Media, 2012.

"Firesign World / Anytown USA." *Chromium Switch*, August 1974.

Fiske, John. *Television Culture*. London: Methuen, 1987.

Flynn, John T. *The Truth about Pearl Harbor*. Glasgow: Strickland, 1945.

Fong-Torres, Ben. "FM Underground Radio: Love for Sale." *Rolling Stone*, April 2, 1970.

———. *Hickory Wind: The Life and Times of Gram Parsons*. New York: Pocket Books, 1991.

———, ed. *The Rolling Stone Rock 'n' Roll Reader*. New York: Bantam, 1974.

———. "Wacco's Circle Is Unbroken." *Rolling Stone*, May 27, 1971.

Ford, Douglas. "Crossroads and Cross-Currents in *Invisible Man*." *Modern Fiction Studies* 45, no. 4 (1999): 887–904.

"Four Hours of Firesign Theatre." *Cornell Daily Sun*, Nov. 19, 1971.

Freberg, Stan. *It Only Hurts When I Laugh*. New York: Times Books, 1988.

———. *Stan Freberg Presents the United States of America*. Vol. 1, *The Early Years*. Capitol SW 1573, 1961, LP.

Frith, Simon. "Rock and the Politics of Memory." In *The 60s without Apology*, edited by Sohnya Sayres, Anders Stephanson, Stanley Aronowitz, and Fredric Jameson, 59–69. Minneapolis: University of Minnesota Press, 1984.

———. *Sound Effects: Youth, Leisure, and the Politics of Rock 'n' Roll*. New York: Pantheon, 1983.

Ganz, Cheryl. *The 1933 Chicago World's Fair: Century of Progress*. Urbana: University of Illinois Press, 2008.

Garson, Barbara. "Luddites in Lordstown." *Harpers*, June 1972.

———. *Macbird!* New York: Grove, 1967.

———. *Macbird! A Recording of the Complete Text of the Play with the Original Cast*. Evergreen Records EVR 004, 1967, LP.

Gebhardt, Steve, dir. *Ten for Two: The John Sinclair Freedom Rally*. Joko Productions, 1971.

Gelatt, Roland. *The Fabulous Phonograph: From Tin Foil to High Fidelity*. Philadelphia: J. B. Lippincott, 1955.

Gendron, Bernard. *Between Montmartre and the Mudd Club: Popular Music and the Avant-Garde*. Chicago: University of Chicago Press, 2002.

Gerber, Steve. "I Think We're All Bozos in This Book." *The Defenders*. Marvel Comics Group 1, no. 34 (April 1976).

Ghelderode, Michel de. "Christopher Columbus (Christopher Colomb)." *Seven Plays*. Translated by George Hauger. New York: Hill and Wang, 1964.

Ginsberg, Allen. "Beatles Essay." In *Deliberate Prose: Selected Essays, 1952– 1995*, edited by Bill Morgan, 456. New York: HarperCollins, 2000.

———. "Wichita Vortex Sutra." *Planet News*. San Francisco: City Lights Books, 1968.

Gitlin, Todd. *The Sixties: Years of Hope, Days of Rage*. Toronto: Bantam, 1987.

———. *The Whole World Is Watching: Mass Media in the Making and Unmaking of the New Left*. Berkeley: University of California Press, 2003.

Glass, Loren. *Counterculture Colophon: Grove Press, the Evergreen Review, and the Incorporation of the Avant-Garde*. Stanford, CA: Stanford University Press, 2013.

Goodyear, Anne Collins. "From Technophilia to Technophobia: The Impact of the Vietnam War on the Reception of 'Art and Technology.'" *Leonardo* 41, no. 2 (2008): 169–73.

Gold, Mike. "Muzak of the Spheres." *Chicago Seed*, Oct. 18–Nov. 1, 1972.

———. "Muzak of the Spheres." *Chicago Seed*, Nov. 13–30, 1972.

Goldstein, Richard. "We Still Need the Beatles, but . . ." In *Read the Beatles: Classic and New Writings on the Beatles, Their Legacy, and Why They Still Matter*, edited by June Skinner Sawyers, 97–101. New York: Penguin, 2006.

Gosling, John. *Waging "The War of the Worlds": A History of the 1938 Radio Broadcast and Resulting Panic.* Jefferson, NC: McFarland, 2009.

Gould, Jonathan. *Can't Buy Me Love: The Beatles, Britain, and America.* New York: Harmony, 2007.

Granata, Charles L. *I Just Wasn't Made for These Times: Brian Wilson and the Making of Pet Sounds.* London: Unanimous, 2003.

Greathouse, Lee. "Firesign Fanatics Freak-Out Tonight." *Stanford Daily*, Feb. 5, 1971.

Greenhalgh, Paul. *Fair World: A History of World's Fairs and Expositions, from London to Shanghai, 1851–2010.* Winterbourne, Berkshire: Papadakis, 2011.

Greenspun, Robert. "Screen: 'Zachariah,' an Odd Western." *New York Times*, Jan. 25, 1971.

Grobe, Christopher. "The Programming Era: The Art of Conversation Design from ELIZA to Alexa." *Post45*, March 27, 2023. https://post45.org/2023/03/the-programming-era.

Gunning, Tom. "The Cinema of Attraction: Early Film, Its Spectator and the Avant-Garde." *Wide Angle* 8 (Fall 1986): 63–70.

Gunther, John. *Inside Europe.* New York: Harper, 1936.

Haggins, Bambi, and Amanda D. Lotz. "At Home on the Cutting Edge." In *The Essential HBO Reader*, edited by Gary R. Edgerton and Jeffrey P. Jones, 151–71. Lexington: University Press of Kentucky, 2008.

Haley, Brian D. "Craig Carpenter and the Neo-Indians of LONAI." *American Indian Quarterly* 42, no. 2 (2018): 215–45.

———. *Hopis and the Counterculture: Traditionalism, Appropriation, and the Birth of a Social Field.* Tucson: University of Arizona Press, forthcoming.

Hallin, Daniel C. *The "Uncensored War": The Media and Vietnam.* New York: Oxford University Press, 1986.

Hamilton, Jack. *Just Around Midnight: Rock and Roll and the Racial Imagination.* Cambridge, MA: Harvard University Press, 2016.

Hammond, Paul, ed. and trans. *The Shadow and Its Shadow: Surrealist Writings on the Cinema.* 3rd ed. San Francisco: City Lights Books, 2000.

Harmetz, Aljean. *The Making of "The Wizard of Oz."* Chicago: Chicago Review Press, 2013.

Harrington, Richard. "WHFS: The End of the Rainbow." *Washington Post*, Jan. 16, 1983.

Hartig, Anthea. "'A Most Advantageous Spot on the Map': Promotion and Popular Culture." In *A Companion to Los Angeles*, edited by William Deverell and Greg Hise, 289–312. Malden, MA: Wiley-Blackwell, 2010.

Havel, Václav. *The Garden Party and Other Plays.* New York: Grove, 1993.

Hayden, Tom. "The CIA's Student-Activism Phase." *The Nation*, Nov. 26, 2014. https://www.thenation.com/article/archive/cias-student-activism-phase/.

Hayes, Johnny. "KRLA Airchecks." April 22, 1967. http://krlabeat.sakionline
.net/airchecks.html.

Heibel, Amy. "LACMA Launches Art + Technology Lab." *LACMA Unframed*,
Nov. 16, 2013. https://unframed.lacma.org/2013/12/10/
lacma-launches-art-technology-lab.

Hepworth, David. *Never a Dull Moment: 1971, the Year That Rock Exploded.*
New York: Henry Holt, 2016.

Herbert, Daniel. *Maverick Movies: New Line Cinema and the Transformation
of American Film.* Oakland: University of California Press, 2024.

Hickey, Andrew. "'Dark Star' by the Grateful Dead." *A History of Rock Music in
500 Songs.* Podcast episode 165. May 20, 2023. https://500songs.com
/podcast/episode-165-dark-star-by-the-grateful-dead/.

———. "'Itchycoo Park' by the Small Faces." *A History of Rock Music in 500
Songs.* Podcast episode 159. Dec. 7, 2022. https://500songs.com/podcast
/episode-159-itchycoo-park-by-the-small-faces/.

Higham, Charles. *Orson Welles: The Rise and Fall of an American Genius.* New
York: St. Martin's, 1985.

Hill, Richard. "Chanting the Square Deific." *Rolling Stone*, Sept. 30, 1971.

Hiss, Tony, and David McClelland. "The Firesign Comics Are Hot Stuff." *New
York Times*, April 20, 1975.

Hochman, Brian. *The Listeners: A History of Wiretapping in the United States.*
Cambridge, MA: Harvard University Press, 2022.

———. *Savage Preservation: The Ethnographic Origins of Modern Media
Technology.* Minneapolis: University of Minnesota Press, 2014.

Hodenfield, Chris. "What Price Ecstasy? On 'Let's Make a Deal' the Curtain Is
the Last Frontier." *Rolling Stone*, June 5, 1975.

Hoeffler, Don C. "Silicon Valley U.S.A." *Electronic News*, Jan. 11, 1971.

Hoffman, Abbie. *Revolution for the Hell of It.* New York: Dial, 1968.

Hollenbach, Lisa. *Poetry FM: American Poetry and Radio Counterculture.* Iowa
City: University of Iowa Press, 2023.

Holzman, Jac, and Gavan Daws. *Follow the Music: The Life and High Times of
Elektra Records in the Great Years of American Pop Culture.* Santa Monica,
CA: First Media, 1998.

Hopkins, Jerry. "Los Angeles Scene: A Special Report." *Rolling Stone*, June 22,
1968.

Horten, Gerd. *Radio Goes to War: The Cultural Politics of Propaganda during
World War II.* Berkeley: University of California Press, 2002.

Hoskyns, Barney. *Waiting for the Sun: Strange Days, Weird Scenes, and the
Sound of Los Angeles.* New York: St. Martin's, 1996.

Howard, David N. *Sonic Alchemy: Visionary Music Producers and Their
Maverick Recordings.* Milwaukee, WI: Hal Leonard, 2004.

Hughes, Ken. *Chasing Shadows: The Nixon Tapes, the Chennault Affair, and the Origins of Watergate.* Charlottesville: University of Virginia Press, 2014.

Hunt, Tim. *The Textuality of Soulwork: Jack Kerouac's Quest for Spontaneous Prose.* Ann Arbor: University of Michigan Press, 2014.

The I Ching, or, Book of Changes. Richard Wilhelm translation, rendered into English by Cary F. Baynes. Foreword by C. G. Jung. 1950. New York: Pantheon, 1961.

Ingham, John, ed. "Art Is Cheese Made Visible: Religious Frenzy with the Firesign Theatre." *Creem*, Oct. 1972.

James, David E. *Rock 'n' Film: Cinema's Dance with Popular Music.* New York: Oxford University Press, 2016.

James, Edwin H. "The Hard Way to Make Money in Television." *Broadcasting*, March 5, 1973.

Jenkins, Henry. *Textual Poachers: Television Fans and Participatory Culture.* Updated ed. New York: Routledge, 2013.

Jessen, Taylor. "An Avant-Garde Religious Radio Happening." In Firesign Theatre, *Duke of Madness Motors* (booklet), 8–32. Canada: Seeland, 2010.

Johnston, Laurie. "Lennon Sees a Wide Impact in Ouster." *New York Times*, April 3, 1973.

Joyce, Don. "Negativland Interview." Interview by Jason Gross. *Perfect Sound Forever.* June 2000. http://www.furious.com/perfect/negativland.html.

Kael, Pauline. "Bonnie and Clyde." *New Yorker*, Oct. 21, 1967.

———. "El Poto-Head Comics." *New Yorker*, Nov. 20, 1971.

Kalat, David. *A Critical History and Filmography of Toho's Godzilla Series.* 2nd ed. Jefferson, NC: McFarland, 2010.

Karnow, Stanley. *Vietnam: A History.* New York: Penguin, 1984.

Karpel, Craig. "George Who for Veep." *Creem*, Nov. 1972.

Katz, Mark. *Groove Music: The Art and Culture of the Hip-Hop DJ.* New York: Oxford University Press, 2012.

Kauffman, L. A. *Direct Action: Protest and the Reinvention of American Radicalism.* London: Verso, 2017.

Kealy, Edward R. "From Craft to Art: The Case of Sound Mixers and Popular Music." In *On Record: Rock, Pop, and the Written Word*, edited by Simon Frith and Andrew Goodwin, 207–20. New York: Pantheon, 1990.

Keane, Damien. "An Ear toward Security: The Princeton Listening Center." *Princeton University Library Chronicle* 77, no. 1 (2009): 45–62.

Keightley, Keir. "Long-Play: Adult-Oriented Popular Music and the Temporal Logics of the Post-War Sound Recording Industry in the USA." *Media, Culture & Society* 26, no. 3 (2004): 375–91.

Keith, Michael C. *Voices in the Purple Haze: Underground Radio and the Sixties.* Westport, CT: Praeger, 1997.

Kelley, Tina. "Ron Ziegler, Press Secretary to Richard Nixon, Is Dead at 63." *New York Times*, Feb. 11, 2003. http://www.nytimes.com/2003/02/11/us /ron-ziegler-press-secretary-to-nixon-is-dead-at-63.html.

Kerouac, Jack. *Visions of Cody*. New York: McGraw Hill, 1972.

Kinks, The. *Arthur (or the Decline and Fall of the British Empire)*. Reprise Records RS 6366, 1969, LP.

Kittler, Friedrich A. *Gramophone, Film, Typewriter*. Translated by Geoffrey Winthrop-Young and Michael Wutz. Stanford, CA: Stanford University Press, 1999.

———. "Playback: A World War History of Radio Drama." Translated by Michael Wutz. In *Operation Valhalla: Writings on War, Weapons, and Media*, edited by Geoffrey Winthrop-Young, Michael Wutz, and Ilinca Iurascu, 91–109. Durham, NC: Duke University Press, 2021.

Koethe, John. "Ninety-Fifth Street." *Poetry*, July/August 2009.

Kozloff, Max. "The Multimillion Dollar Art Boondoggle." *Artforum*, Oct. 1971.

KPFK Folio 8, no. 9 (March 1967). https://archive.org/details/ marcfolio1311967kpfkrich.

Kramer, Michael J. *The Republic of Rock: Music and Citizenship in the Sixties Counterculture*. New York: Oxford University Press, 2013.

Krieger, Susan. *Hip Capitalism*. Beverly Hills, CA: Sage, 1979.

Kruse, Kevin M. *One Nation under God: How Corporate America Invented Christian America*. New York: Basic Books, 2015.

Kruse, Michael. "The Escalator Ride That Changed America." *Politico*, June 16, 2019. https://www.politico.eu/article/ the-escalator-ride-that-changed-america-donald-trump-presidential-bid.

Kubernik, Harvey, and Kenneth Kubernik. *A Perfect Haze: The Illustrated History of the Monterey International Pop Festival*. Solana Beach, CA: Santa Monica Press, 2011.

Kurlansky, Mark. *1968: The Year That Rocked the World*. New York: Random House, 2005.

Kyrou, Ado. *Le Surréalisme au Cinéma*. La Fléche, France: Le Terrain Vague, 1963.

Lambert, Philip. *Good Vibrations: Brian Wilson and the Beach Boys in Critical Perspective*. Ann Arbor: University of Michigan Press, 2016.

Land, Jeff. *Active Radio: Pacifica's Brash Experiment*. Minneapolis: University of Minnesota Press, 1999.

LaRene, Deday. "The Walrus Was Paul?" *Creem*, Nov. 1969.

Lask, Thomas. "Recordings: Comedians, Savage, Friendly and Otherwise." *New York Times*, August 18, 1968.

Le Blanc, Jerry, and Ivor Davis. *5 to Die*. Los Angeles: Holloway House, 1970.

Lee, Hermione. *Tom Stoppard: A Life*. New York: Alfred A. Knopf, 2021.

Lennon, John. "The Goon Show Scripts." *New York Times*, Sept. 30, 1973.

Lesnick, Henry, ed. *Guerilla Street Theater*. New York: Bard/Avon, 1973.

Lewisohn, Mark. *The Beatles Recording Sessions*. New York: Harmony, 1988.

Linder, Michael. "A Salute to Columbia Square." KNX-AM, 2005. https://www
.linder.tv/mp3/Columbia_Square.mp3.

Lott, Eric. *Black Mirror: The Cultural Contradictions of American Racism*.
Cambridge, MA: Belknap Press of Harvard University Press, 2017.

———. *Love and Theft: Blackface Minstrelsy and the American Working Class*.
New York: Oxford University Press, 1993.

Lotz, Amanda D. *We Now Disrupt This Broadcast: How Cable Transformed
Television and the Internet Revolutionized It All*. Cambridge, MA: MIT
Press, 2018.

Malloch, William. Interview by Chris Palladino, June 1, 1995. Firesign Theatre
Collection, National Audio-Visual Conservation Center, Library of Congress.

Malone, Eric. "Postmark: Deep Space." *Four-Alarm FIRESIGNal* 22, Dec. 1993.

"Manson Is Music 'Addict': Theory Links Beatle Album to Murders." *Los Angeles
Times*, Feb. 6, 1970.

Marcus, Greil. "The Beatles." In *The Rolling Stone Illustrated History of Rock &
Roll*, edited by Jim Miller, 177–89. Rev. ed. New York: Random House, 1980.

———. "The Firesign Theatre." In *The Rolling Stone Record Guide*, edited by
Dave Marsh and John Swenson, 129–30. New York: Random House/Rolling
Stone Press, 1979.

———. *Lipstick Traces: A Secret History of the Twentieth Century*. Cambridge,
MA: Harvard University Press, 1989.

———. *Mystery Train: Images of America in Rock 'n' Roll Music*. 4th rev. ed.
1975. New York: Plume, 1997.

———. "Rock-a-Hula Clarified." *Creem*, June 1971.

Markle, Seth M. "The Hip-Hop DJ as Black Archaeologist: Madlib's *Beat
Konducta in Africa* and the Politics of Memory." In *Politics of an Anticolo-
nial Archive*, edited by Shiera S. el-Malik and Isaac A. Kamola, 207–29.
London: Rowman & Littlefield, 2018.

Marmorstein, Gary. *The Label: The Story of Columbia Records*. New York:
Thunder's Mouth, 2007.

Marsh, Dave. Review of *I Think We're All Bozos on This Bus*, by the Firesign
Theatre. *Creem*, Nov. 1971.

Marsh, Dave, and Kevin Stein. *The Book of Rock Lists*. New York: Dell/Rolling
Stone Press, 1981.

Mason, Nick. *Inside Out: A Personal History of Pink Floyd*. Edited by Philip
Dodd. San Francisco: Chronicle, 2017.

McCarthy, Anna. *Ambient Television: Visual Culture and Public Space*.
Durham, NC: Duke University Press, 2001.

McCarthy, David Michael. "On the Appearance of the Comedy LP, 1957–1973."
PhD diss., City University of New York, 2016.

McClay, Bob. "Industry's All-Stereo Push Puts the Needle in Consumer Instead of Inbetween the Grooves." *Rolling Stone*, July 6, 1968.

McGuinn, Roger, and David Crosby. "'Interview' for DJs." On *Fifth Dimension*, by the Byrds. Columbia Legacy 483707 2, 1996, CD.

McKenzie, D. F. *Bibliography and the Sociology of Texts*. London: British Library, 1986.

McKinney, Devin. *Magic Circles: The Beatles in Dream and History*. Cambridge, MA: Harvard University Press, 2003.

McLeod, Kembrew. *Pranksters: Making Mischief in the Modern World*. New York: New York University Press, 2014.

McLuhan, Marshall. *Understanding Media: The Extensions of Man*. New York: Signet, 1964.

McLuhan, Marshall, and Quentin Fiore. *War and Peace in the Global Village: An Inventory of Some of the Current Spastic Situations That Could Be Eliminated by More Feedforward*. New York: McGraw-Hill, 1968.

McMurria, James. *Republic on the Wire: Cable Television, Pluralism, and the Politics of New Technologies, 1948–1984*. New Brunswick, NJ: Rutgers University Press, 2017.

Mead, Margaret. "'As Significant as the Invention of Drama or the Novel': A Famed Anthropologist Takes a Careful Look at *An American Family*." *TV Guide*, Jan. 6, 1973.

Merline, David. "The Firesign Theatre: A Young Person's Guide." *Motorbooty* 6 (Fall 1992).

Milligan, Spike. *Adolf Hitler: My Part in His Downfall*. New York: Harper's Magazine Press, 1974.

Mills, Paul. "The Firesign Theatre." *Fusion*. Feb. 5, 1971.

Milner, Greg. *Perfecting Sound Forever: An Aural History of Recorded Music*. New York: Faber & Faber, 2009.

Monaco, Paul. *History of the American Cinema*. Vol. 8, *The Sixties, 1960–1969*. Edited by Charles Harpole. Berkeley: University of California Press, 2001.

Monroe, Larry, and Jim Dulzo. Quoted in David Fenton, "'We're Selling Too Much Death': Mutiny at WABX." *Ann Arbor Sun*, June 23–July 7, 1972.

Mullen, Megan. *The Rise of Cable Programming in the United States: Revolution or Evolution?* Austin: University of Texas Press, 2003.

Ndiaye, Noémie. *Scripts of Blackness: Early Modern Performance Culture and the Making of Race*. Philadelphia: University of Pennsylvania Press, 2022.

Neer, Richard. *FM: The Rise and Fall of Rock Radio*. New York: Villard, 2001.

Neill. Review of *I Think We're All Bozos on This Bus*, by the Firesign Theatre. *International Times*, Jan. 27, 1972.

"New Recording Facilities for Technical Operations." *Columbia Record*, April-May 1960.

Nisker, Wes "Scoop." In *If You Don't Like the News . . . Go Out and Make Some of Your Own!* Berkeley: Ten Speed, 1994.

Nuts to You. Insert in *Motorbooty* 7 (Fall 1994).

"'Obscene' Airwaves." *Daily Texan,* Oct. 25, 1972.

Official Guide Book of the Fair 1933. Chicago: Century of Progress, 1933.

O'Hagan, Sean. "De La Soul: Three Feet High and Rising (Big Life LP/Cassette /CD)." *New Musical Express,* March 18, 1989. *Rock's Backpages.* http://www .rocksbackpages.com/Library/Article/de-la-soul-ithree-feet-high-and -risingi-big-life-lpcassettecd.

Ondaatje, Michael, and Walter Murch. *The Conversations: Walter Murch and the Art of Editing Film.* New York: Borzoi, 2002.

Ong, Walter J. *Orality and Literacy: The Technologizing of the Word.* 1982. London: Routledge, 2002.

Ossman, David. *Dr. Firesign's Follies: Radio, Comedy, Mystery, History.* Albany, GA: BearManor Media, 2008.

———. Interview by Kurt Ericson, May 7, 2015. https://archive.org/details /OssmanChatInterview20150507.

———. Liner notes for Firesign Theatre, *How Can You Be in Two Places at Once When You're Not Anywhere at All.* Mobile Fidelity Sound Lab MFCD 834, 1988. Originally released in 1969 by Columbia Records.

———. Liner notes for Firesign Theatre, *Waiting for the Electrician or Someone Like Him.* Mobile Fidelity Sound Lab MFCD 762, 1992. Originally released in 1968 by Columbia Records.

———. "The Martian Space Party Diary: A Chronicle of the Firesign Theatre, January–June 1972." https://www.firesigntheatre.com/martian/index.html. Accessed on Wayback Machine, August 9, 2023.

———. "A Memoir of the Renaissance Pleasure Faire." Unpublished manuscript, 2016.

———. "The Odyssey of Me and Norman Corwin." In *Anatomy of Sound: Norman Corwin and Media Authorship,* edited by Jacob Smith and Neil Verma, 211–32. Oakland: University of California Press, 2016.

———. *The Sullen Art: Recording the Revolution in American Poetry.* Rev. ed. Toledo, OH: University of Toledo Press, 2016.

———. "Zack: A Look Back." *Four-Alarm FIRESIGNal* 3, August 1984.

Oxford English Dictionary, s.v. "stratification (n.)," July 2023. https://doi.org/10 .1093/OED/9107586971.

Paget, Karen M. *Patriotic Betrayal: The Inside Story of the CIA's Secret Campaign to Enroll American Students in the Crusade against Communism.* New Haven, CT: Yale University Press, 2015.

Pansey, Susan L. "A Vest Has No Sleeves." *Village Voice,* June 18, 1970.

Parks, Lisa. *Cultures in Orbit: Satellites and the Televisual.* Durham, NC: Duke University Press, 2005.

Pas, Niek. "Mediatization of the Provos: From a Local Movement to a European Phenomenon." In *Between Prague Spring and French May: Opposition and Revolt in Europe, 1960–1980,* edited by Martin Klimke, Jacco Pekelder, and Joachim Scharloth, 157–76. New York: Berghahn, 2011.

Peck, Ellen. "The Column." *Dubuque Telegraph Herald,* Nov. 24, 1971.

Perchard, Tom. "Hip Hop Samples Jazz: Dynamics of Cultural Memory and Musical Tradition in the African American 1990s." *American Music* 29, no. 3 (2011): 277–307.

Perlstein, Rick. *Nixonland: The Rise of a President and the Fracturing of America.* New York: Scribner, 2008.

Perrin, Tom. "The Great American Novel in the 1970s." In *American Literature in Transition, 1970–1980,* edited by Kirk Curnutt, 196–209. Cambridge: Cambridge University Press, 2018.

Perry, Charles. "Is This Any Way to Run the Army—Stoned?" *Rolling Stone,* Nov. 9, 1968.

"Peter Bergman's 'Oz' . . . Happening." *KRLA Beat,* April 22, 1967.

Phan, Thao. "Amazon Echo and the Aesthetics of Whiteness." *Catalyst: Feminism, Theory, Technoscience* 5, no. 1 (2019): 1–37.

Piekut, Benjamin. *Henry Cow: The World Is a Problem.* Durham, NC: Duke University Press, 2019.

Pierce, Dave. *Riding on the Ether Express: A Memoir of 1960s Los Angeles, the Rise of Freeform Underground Radio, and the Legendary KPPC-FM.* Lafayette: Center for Louisiana Studies, University of Louisiana at Lafayette, 2008.

Pinch, Trevor. *Analog Days: The Invention and Impact of the Moog Synthesizer.* Cambridge, MA: Harvard University Press, 2002.

———. "Stanley Milgram and the Sonic Imaginary." Unpublished manuscript, 2016.

Plant, Robert. Interview. *Friends,* Feb. 20, 1970.

"Pop Records: Moguls, Money & Monsters." *Time,* Feb. 12, 1973.

Povey, Glenn. *Echoes: The Complete History of Pink Floyd.* Chicago: Chicago Review Press, 2010.

Proctor, Phil. Interview by Jason Klamm. "The History of Firesign, Part 5: *I Think We're [All] Bozos on This Bus.*" *Comedy on Vinyl.* Podcast episode 269. August 9, 2018. https://stolendress.com/comedyonvinyl/episode-269-the-history-of-firesign-part-5-i-think-were-all-bozos-on-this-bus/.

Proctor, Phil, and Brad Schreiber. *Where's My Fortune Cookie? (My Psychic, Psurrealistic Pstory).* Studio City, CA: Parallel Universe, 2017.

"Programming Aids." *Billboard,* Nov. 2, 1968.

"Provos: Blasting Back at the Brave New World." *Los Angeles Free Press,* Nov. 25, 1966.

Radway, Janice. "Zines, Half-Lives, and Afterlives: On the Temporalities of Social and Political Change." *PMLA* 126, no. 1 (Jan. 2011): 140–50.

"Random Notes." *Rolling Stone*, Oct. 1972.

Ray, Robert B. "Tracking." In *Present Tense: Rock & Roll and Culture*, edited by Anthony DeCurtis, 135–48. Durham, NC: Duke University Press, 1992.

Reeuwijk, Dick P. J. van. *Damsterdamse Extremisten*. Amsterdam: Uitgeverij de Bezige Bij, 1965.

Reeve, Andru J. *Turn Me On, Dead Man: The Beatles and the 'Paul-Is-Dead' Hoax*. Bloomington, IN: AuthorHouse, 2004.

Reynolds, Simon. *Rip It Up and Start Again: Postpunk 1978–1984*. New York: Penguin, 2005.

Richard, Paul. "'Cheese' Melts Off Air." *Washington Post*, June 2, 1970.

Riggs, Elayne. "In My Ears and in My Eyes (Part 2)." *ComicMix*, March 26, 2008. https://mix.malibulist.com/2008/03/26/in-my-ears-and-in-my-eyes-part-2-by-elayne-riggs.

———. "Postmark: Deep Space." *Four-Alarm FIRESIGNal* 9, August 1986.

Ritter, Niles. "Hamburger All Over the Info.Highway." *Four-Alarm FIRESIGNal* 24, August 1994.

Robbins, Trina. *Last Girl Standing*. Seattle, WA: Fantagraphics, 2017.

Roberts, Howard. *Antelope Freeway*. Impulse! AS 9207, 1971, LP.

Roberts, Jem. *Fab Fools: The Last Untold Story of the Beatles*. Foreword by Kevin Eldon. Cardiff: Candy Jar, 2021.

Roberts, Mervyn Edwin, III. *The Psychological War for Vietnam, 1960–1968*. Lawrence: University Press of Kansas, 2018.

"Rock-a-Rama." *Creem*, Feb. 1973.

Rogan, Johnny. *The Byrds: Timeless Flight Revisited: The Sequel*. London: Rogan House, 1997.

Roose, Kevin. "Bing's A.I. Chat: 'I Want to Be Alive 😈.'" *New York Times*, Feb. 17, 2023. https://www.nytimes.com/2023/02/16/technology/bing-chatbot-transcript.html.

Rose, Tricia. *Black Noise: Rap Music and Black Culture in Contemporary America*. Hanover, NH: Wesleyan University Press, 1994.

Roszak, Theodore. *The Making of a Counterculture: Reflections on the Technocratic Society and Its Youthful Opposition*. 1969. Berkeley: University of California Press, 1995.

Rubin, Mike, Rob Michaels, David Merline, Dan Rice, and Barry Henssler. "Rock Lit: A *Motorbooty* Survey." *Motorbooty* 7 (Fall 1994).

Rubin, Rachel Lee. *Well Met: Renaissance Faires and the American Counterculture*. New York: New York University Press, 2012.

Rudd, Mark. *Underground: My Life with SDS and the Weathermen*. New York: William Morrow, 2009.

Saerchinger, César. *Hello, America! Radio Adventures in Europe*. Boston: Houghton Mifflin, 1938.

Sagittarius. "Hotel Indiscreet." Columbia, 4-44289, 1967, 7″.

———. "My World Fell Down." Columbia, 4-44163, 1967, 7″.

Saunders, Frances Stonor. *The Cultural Cold War: The CIA and the World of Arts and Letters*. New York: New Press, 1999.

Savage, Jon. *1966: The Year the Decade Exploded*. London: Faber & Faber, 2015.

Schafer, Murray R. "The Music of the Environment." In *Audio Culture: Readings in Modern Music*, edited by Christopher Cox and Daniel Warner, 31–41. Rev. ed. New York: Bloomsbury, 2017.

Schmidt Horning, Susan. *Chasing Sound: Technology, Culture, and the Art of Studio Recording from Edison to the LP*. Baltimore: Johns Hopkins University Press, 2013.

Scoppa, Bud. "Los Lobos: 30 Years of Eclectic Rock." *Paste*, June 1, 2004. *Rock's Backpages*. http://www.rocksbackpages.com/Library/Article/los-lobos-30-years-of-eclectic-rock.

Scott, Ken. "Ken Scott on the Beatles White Album." Interview by Jason Barnard. *The Strange Brew*, Sept. 30, 2018. https://thestrangebrew.co.uk/ken-scott-on-the-beatles-white-album.

Sergi, Gianluca. *The Dolby Era: Film Sound in Contemporary Hollywood*. Manchester: Manchester University Press, 2004.

Seydor, Paul. *The Authentic Death and Contentious Afterlife of Pat Garrett and Billy the Kid: The Untold Story of Peckinpah's Last Western Film*. Evanston, IL: Northwestern University Press, 2015.

Shamberg, Michael, and Raindance Corporation. *Guerrilla Television*. New York: Henry Holt, 1971.

Sheffield, Rob. *Dreaming the Beatles: The Love Story of One Band and the Whole World*. New York: William Morrow, 2017.

Shoe, Merril. "On the Campoon Trail." *Berkeley Barb*, Nov. 21–27, 1975.

Simon, Ed. "Who Still Needs the Carnivalesque?" *The Baffler*, Nov. 7, 2022. https://thebaffler.com/latest/who-still-needs-the-carnivalesque-simon.

Simon, John. *Truth, Lies & Hearsay: A Memoir of a Life in and out of Rock and Roll*. Self-published, 2019.

Simpson, Christopher. *Science of Coercion: Communication Research and Psychological Warfare, 1945–1960*. New York: Oxford University Press, 1994.

Simpson, Kim. *Early '70s Radio: The American Format Revolution*. New York: Continuum, 2011.

Sinykin, Dan. "Charles Manson and the Apocalypse to Come." *Los Angeles Review of Books Blog*, Nov. 24, 2017. https://blog.lareviewofbooks.org/essays/charles-manson-apocalypse-come.

Situationist International. *Ten Days That Shook the University*. 1966. https://rozsixties.unl.edu/items/show/688.

Sloan Commission on Cable Communications. *On the Cable: The Television of Abundance*. New York: McGraw-Hill, 1971.

Smith, Jacob. *Spoken Word: Postwar American Phonograph Cultures*. Berkeley: University of California Press, 2011.

Smith, Sherry L. *Hippies, Indians, and the Fight for Red Power*. Oxford: Oxford University Press, 2012.

Snyder, Alvin A. *Warriors of Disinformation: American Propaganda, Soviet Lies, and the Winning of the Cold War, an Insider's Account*. New York: Arcade, 1995.

Sontag, Susan. "Happenings: An Art of Radical Juxtaposition." In *Against Interpretation*, 263–74. New York: Farrar, Straus & Giroux, 1966.

Spigel, Lynn. *TV by Design: Modern Art and the Rise of Network Television*. Chicago: University of Chicago Press, 2008.

St. Clair Smith, Douglass, dir. *Let's Visit the World of the Future*. Bong Bulldada Time Control, 1973.

Stadler, Gustavus. "'My Wife': The Tape Recorder and Warhol's Queer Ways of Listening." *Criticism* 56, no. 3 (2014): 425–56.

Steinitz, Kate. Interview by Philip Proctor and David Ossman. *Radio Free Oz Elements:* "Summer of Love." 1967. Firesign Theatre Collection, National Audio-Visual Conservation Center, Library of Congress.

Stern, Sol. "A Short Account of International Student Politics and the Cold War with Particular Reference to the NSA, CIA, etc." *Ramparts*, March 1967.

Sterne, Jonathan. *The Audible Past: Cultural Origins of Sound Reproduction*. Durham, NC: Duke University Press, 2003.

———. "The Stereophonic Spaces of Soundscape." In *Living Stereo: Histories and Cultures of Multichannel Sound*, edited by Paul Théberge, Kyle Devine, and Tom Everrett, 64–84. New York: Bloomsbury, 2015.

Sterne, Jonathan, and Mehak Sawheny. "The Acousmatic Question and the Will to Datafy: Otter.ai, Low-Resource Languages, and the Politics of Machine Learning." *Kalfou* 9, no. 2 (2022): 288–306.

Steyerl, Hito. "The Spam of the Earth: Withdraw from Representation." In *Hito Steyerl: The Wretched of the Screen*, 160–75. Berlin: Sternberg, 2012.

Stockwell, Norman. "Former Breakfast Special Host Chris Ward Passes." WORT Eighty Nine.Nine FM, Oct. 23, 2015. https://www.wortfm.org/former-breakfast-special-host-chris-ward-passes/.

Stoever, Jennifer Lynn. *The Sonic Color Line: Race and the Cultural Politics of Listening*. New York: New York University Press, 2016.

Streeter, Thomas. "Blue Skies and Strange Bedfellows: The Discourse of Cable Television." In *The Revolution Wasn't Televised*, edited by Lynn Spigel and Michael Curtin, 221–42. New York: Routledge, 1997.

———. *Selling the Air: A Critique of the Policy of Commercial Broadcasting in the United States*. Chicago: University of Chicago Press, 1996.

Stuart, Chad, and Jeremy Clyde. *Of Cabbages and Kings*. Columbia, CS 9471, 1967, LP.

Tarnoff, Ben. "Weizenbaum's Nightmares: How the Inventor of the First Chatbot Turned Against AI." *The Guardian*, July 25, 2023. https://www .theguardian.com/technology/2023/jul/25/joseph-weizenbaum-inventor-eliza-chatbot-turned-against-artificial-intelligence-ai.

Tate, Greg, ed. *Everything but the Burden: What White People Are Taking from Black Culture*. New York: Harlem Moon/Broadway, 2003.

———. *Flyboy in the Buttermilk: Essays on Contemporary America*. New York: Simon & Schuster, 1992.

Taylor, Adam. "7 Times the Onion Was Lost in Translation." *Washington Post*, June 15, 2015. https://www.washingtonpost.com/news/worldviews/wp/2015 /06/02/7-times-the-onion-was-lost-in-translation/.

Terry, Wallace, ed. *Bloods, an Oral History of the Vietnam War*. New York: Random House, 1984.

Théberge, Paul, Kyle Devine, and Tom Everrett. Introduction to *Living Stereo: Histories and Cultures of Multichannel Sound*, edited by Paul Théberge, Kyle Devine, and Tom Everrett, 1–36. New York: Bloomsbury, 2017.

Thompson, Dorothy. "The Great War of Words." *Saturday Evening Post*, Dec. 1, 1934.

———. "On the Record: Mr. Welles and Mass Delusion." *New York Herald Tribune*, Nov. 2, 1938.

Thompson, Hunter S. "Memo from the Sports Desk: The So-Called 'Jesus Freak' Scare." *Rolling Stone*, Sept. 2, 1971.

———. "Memoirs of a Wretched Weekend in Washington." In *The Great Shark Hunt: Strange Tales from a Strange Time*, 177–82. New York: Warner, 1979.

Tiegel, Eliot. "School of Relevant Humor Makes the Campus a Fun Place to Study." *Billboard*, March 1971.

Tilly, Charles. "Collective Violence in European Perspective." In *Violence in America: Historical and Comparative Perspectives: A Report to the National Commission on the Causes and Prevention of Violence*, edited by Hugh Davis Graham and Ted Robert Gurr, 5–34. Washington, DC: US Government Printing Office, 1969.

Tingen, Paul. *Miles Beyond: The Electric Explorations of Miles Davis, 1967–1991*. New York: Billboard Books, 2001.

Tirebiter, Mrs. George. "$5 Favorite." *Chicago Tribune*, Jan. 13, 1973.

Tomlinson, Gary. "Cultural Dialogics and Jazz: A White Historian Signifies." Reprinted in "The Best of BMRJ." Supplement, *Black Music Research Journal* 22 (2002): 71–105. Originally published in vol. 11, no. 2 (1991): 229–64.

"Top LP's." *Billboard*, Oct. 18, 1969.

"Top LP's." *Billboard*, Oct. 30, 1971.

Torey, Sal. "Melted Cheese." *Quicksilver Times*, May 19, 1970.

Toubkin, Chrissie, and Phil Vellender. "Lone Survivor's Guide to Firesign Theatre." *Trailing Clouds of Glory* 1 (1974).

Tozer, Lowell. "A Century of Progress, 1833–1933: Technology's Triumph over Man." *American Quarterly* 4, no. 1 (1952): 78–81.

Tris, Laurie. "Firesign Theatre at the Ash Grove." *Los Angeles Free Press*, Nov. 20, 1970.

Tuchman, Maurice. *A Report on the Art and Technology Program of the Los Angeles County Museum of Art, 1967–1971*. Los Angeles: Los Angeles County Museum of Art, 1971.

Turner, Fred. *The Democratic Surround: Multimedia and American Liberalism from World War II to the Psychedelic Sixties*. Chicago: University of Chicago Press, 2013.

Turner, Steve. *Beatles '66: The Revolutionary Year*. New York: Ecco, 2016.

Tytell, John. *The Living Theatre: Art, Exile, and Outrage*. New York: Grove, 1995.

Uhelszki, Jaan. "In the Beginning: 1969–1973." *Creem*, June 1, 2022.

Underwood, Tom B., et al. *Cherokee Legends and the Trail of Tears: From the Nineteenth Annual Report of the Bureau of American Ethnology*. Knoxville: S. B. Newman, 1956.

United States Dept. of Justice. Peter Bergman file obtained under provisions of the Freedom of Information Act. Assorted documents dated August 14, 1962, to Jan. 31, 1963. Internal case file no. 100-HQ-439079.

Uricchio, William. "Television's Next Generation: Technology / Interface Culture / Flow." In *Television after TV: Essays on a Medium in Transition*, edited by Lynn Spigel and Jan Olsson, 163–82. Durham, NC: Duke University Press, 2004.

Ventham, Maxine. *Spike Milligan: His Part in Our Lives*. London: Robson, 2002.

Verma, Neil. *Theater of the Mind: Imagination, Aesthetics, and American Radio Drama*. Chicago: University of Chicago Press, 2012.

Villarejo, Amy. *Ethereal Queer: Television, Historicity, Desire*. Durham, NC: Duke University Press, 2014.

Vizenor, Gerald Robert. *Manifest Manners: Postindian Warriors of Survivance*. Hanover, NH: Wesleyan University Press, 1994.

Wald, Elijah. *How the Beatles Destroyed Rock 'n' Roll: An Alternative History of American Popular Music*. Oxford: Oxford University Press, 2009.

Wald, Gayle. *It's Been Beautiful: Soul! and Black Power Television*. Durham, NC: Duke University Press, 2015.

Walker, Jesse. *Rebels on the Air: An Alternative History of Radio in America*. New York: New York University Press, 2001.

Walker, Michael. *What You Want Is in the Limo: On the Road with Led Zeppelin, Alice Cooper, and the Who in 1973, the Year the Sixties Died and the Modern Rock Star Was Born*. New York: Spiegel and Grau, 2013.

Walls, Richard C. Review of *How Can You Be in Two Places at Once When You're Not Anywhere at All*, by the Firesign Theatre. *Creem*, Nov. 1969.

———. Review of *In the Next World, You're on Your Own*, by Firesign Theatre and *What This Country Needs*, by Proctor and Bergman. *Creem*, Jan. 1976.

Ward, Chris. "Postmark: Deep Space." Letter. *Four-Alarm FIRESIGNal* 7, Dec. 1985.

Ward, Ed. Review of *Waiting for the Electrician or Someone Like Him* and *How Can You Be in Two Places at Once When You're Not Anywhere at All* by the Firesign Theatre. *Rolling Stone*, Nov. 27, 1969.

———. "Through Tirebiter's Television." *Rolling Stone*, Oct. 15, 1970.

———. "Why Do Kids Love These Four Zany Guys?" *New York Times*, Feb. 6, 1972.

Wardrip-Fruin, Noah. *Expressive Processing: Digital Fictions, Computer Games, and Software Studies*. Cambridge, MA: MIT Press, 2009.

Weheliye, Alexander G. *Phonographies: Grooves in Sonic Afro-Modernity*. Durham, NC: Duke University Press, 2005.

Weisbard, Eric. *Top 40 Democracy: The Rival Mainstreams of American Music*. Chicago: University of Chicago Press, 2014.

Weizenbaum, Joseph. *Computer Power and Human Reason: From Judgment to Calculation*. San Francisco: W. H. Freeman, 1976.

———. "Contextual Understanding by Computers." *Computational Linguistics* 10, no. 8 (August 1967): 474–80.

———. "ELIZA—A Computer Program for the Study of Natural Language Communication between Man and Machine." *Computational Linguistics* 9, no. 1 (Jan. 1966): 36–45.

Who, The. *The Who Sell Out*. Decca, DL 74950, 1967, LP.

Wiebel, Frederick C. *Backwards into the Future: The Recorded History of the Firesign Theatre*. Edited by Gregory J. M. Catsos. Boalsburg, PA: BearManor Media, 2006.

Wilcock, John. "A Love-In Inventory." *Los Angeles Free Press*, March 31, 1967.

Williams, Raymond. "Gravity's Python." In *What I Came to Say*, 108–12. London: Hutchinson Radius, 1989.

———. *Raymond Williams on Television: Selected Writings*. Edited by Alan O'Connor. London: Routledge, 1989.

———. *Television: Technology and Cultural Form*. 3rd ed. 1974. London: Routledge, 2003.

Willis, Ellen. "Creedence Clearwater Revival." In *The Rolling Stone Illustrated History of Rock & Roll*, edited by Jim Miller, 324–26. Rev. ed. New York: Random House, 1980.

———. "Herbert Marcuse, 1898–1979." In *Beginning to See the Light*, 141–44. Minneapolis: University of Minnesota Press, 2012.

———. "Records: Rock, Etc.: Pop Ecumenicism." *New Yorker*, May 4, 1968.

———. "Records: Rock, Etc." *New Yorker*, August 10, 1968.

———. "Rock, Etc." *New Yorker*, July 12, 1969.

———. "Rock, Etc." *New Yorker*, August 12, 1972.

———. "Rock, Etc.: Into the Seventies, for Real." *New Yorker*, Dec. 2, 1972.

Winthrop-Young, Geoffrey. "Drill and Distraction in the Yellow Submarine: On the Dominance of War in Friedrich Kittler's Media Theory." *Critical Inquiry* 28, no. 4 (2002): 825–54.

Womack, Kenneth. *Solid State: The Story of "Abbey Road" and the End of the Beatles*. Ithaca, NY: Cornell University Press, 2019.

Wright, Barton. *Clowns of the Hopi: Tradition Keepers and Delight Makers*. Flagstaff, AR: Northland, 1994.

Young, J. R. Review of *Dear Friends*, by the Firesign Theatre. *Rolling Stone*, March 30, 1972.

Zinn, Howard. *A People's History of the United States: 1492–Present*. New York: Harper & Row, 1980.

Zuboff, Shoshana. *The Age of Surveillance Capitalism: The Fight for a Human Future at the New Frontier of Power*. New York: Public Affairs, 2019.

Index

"It's Up to You" (Steinski and Mass Media, 1991), 214–15

Jackson State University, Black students killed at, 91, 116
Jagger, Mick, 70
James, David E., 98
James Gang (band), 98, 100; *Rides Again*, 247n30
Japanese-American internment camps, 190
Jarry, Alfred (*Ubu Cocu*), 24, 36, 127
jazz, xi, 24, 33, 37, 44–45, 76, 77, 216, 235n8
J. Dilla, 215, 216, 218
Jefferson Airplane (band), 47
Jel, DJ, 215
Jenkins, Henry, 205–6
Jeru the Damaga, 215
Jesus Christ Superstar (rock opera, 1970, 1973), 101
"Jesus freaks," 117, 118
Jet Propulsion Laboratory, 159
Jimi Hendrix Experience (band): *Are You Experienced* (1967), 18; "Axis: Bold as Love" (1967), 66; *Electric Ladyland* (1968), 77; *Smash Hits* (1968), 248n30. *See also* Hendrix, Jimi
J-Men Forever (Proctor and Bergman, 1979), 255n6
Jobs, Steve, 161
Jodorowsky, Alejandro, 99
John Sinclair Freedom Rally (Ann Arbor, 1971), 137
Johnson, Lyndon (LBJ), 22, 51, 73
Jones, Brian, 67
Jones, Elvin, 98
Joyce, James, 92; *Finnegans Wake*, 93, 95; *Ulysses*, 64*fig.*, 65, 93, 95
J. Rocc, 215
JUSPAO (Joint United States Public Affairs Office), 51

Kael, Pauline, 50, 52, 99, 247n27
Kafka, Franz, 3, 18, 33
Kane, Big Daddy, 15
Kaprow, Alan, 92, 126
Karamu House (Cleveland theater), 5
Karloff, Boris, 248n38
Kaye, Lenny, 238n64
KCRW-FM radio (Santa Monica), 54
Keaton, Buster, 127, 217
Kennedy, John F., 22, 51
Kent State killings (1970), xii, 114, 116,

126; Devo founder's memory of, 127–28; mediated reference to, 124; protests against Cambodia invasion and, 91, 137; surveillance state and, 131
Kentucky Fried Movie (1977), 165
Kerouac, Jack, 20
Kesey, Ken, 47, 253n32
Kitaj, R. B. (*Lives of the Engineers*), 159
Kittler, Friedrich, 65–67, 68, 83, 236n14
Klein, Allen, 99
KMET-FM radio station (Los Angeles), 77, 78*tab.*, 80, 244n104
KMPX-FM radio station (San Francisco), 76, 257n5
KNIX-FM radio station (Phoenix), 77
Knievel, Evel, 189
Koger, Marijke, 6, 247n25
Kovacs, Ernie, xvii, 167
Kozloff, Max, 159
KPFA-FM radio station (Berkeley), 2, 212
KPFK-FM radio station (Los Angeles), 1, 2, 7, 44, 123, 165, 169, 173*fig.*, 204; *Bald Soprano* (Ionesco, performed 1967), 3; *Christopher Columbus* (Ghelderode, performed 1966), 3, 34; *Dear Friends* program, 77, 133–34; documentaries on Hopi Indians, 3, 22, 29; Firesign Theatre radio programs on, 78*tab.*; fundraising marathons, 4, 8; *The Goon Show* rebroadcast on, 68; *Let's Eat* broadcasts, 171–72; listener responses to put-ons, 50; *Radio Free Oz*, 4, 21–22; Renaissance Pleasure Faire and, 46; "Sound '68" benefit, 170; standard programming of, 2, 22; theater of the absurd and, 3
KPPC-FM radio station (Pasadena), 49, 50, 76, 77, 123; Firesign Theatre radio programs on, 78*tab.*; mass political firings at, 80, 244n112. See also *Radio Hour Hour*
Kramer, Michael, 241n58
Krause, Alison, 127
Krekorian, Kirk, 97
Kristofferson, Kris, 99
KRLA-AM radio station (Los Angeles), 40, 75, 76, 78*tab.*, 80, 244n104
Krofft, Sid and Marty, 159
KSAN-FM radio station (San Francisco), 73, 74, 84; *Berkeley Barb* ad for, 74*fig.*; political firing of Roland Young from, 79, 116, 142
Kuhlman, Kathryn, 117
Kupferberg, Tuli, 2

television, xvi, 93–96, 111, 246n7; as Adorno's "dreamless dream," 188; ambient television, 189, 198; broadcasting as mobile
privatization, 130; changing technologies
and culture of, 163, 164; channel-
switching as media collage, 164; demonstrated at World's Fair (New York, 1939),
163; game shows, 34, 132, 189, 191,
195–96; midnight movies on, 96; Nixon's
"television diplomacy," 174; non-English-
language broadcasting in Los Angeles,
176; nostalgia for world before television,
129; reality TV, xii, 164, 189, 194–95;
reruns, 164, 185–89; simultaneous event
reception through, 199; soap operas, 7,
88, 189, 191; television screen as security
monitor, 199–200; televisual "flow" thesis, 95, 187–88, 246n15; US culture saturated by, 163
Ten Days That Shook the University (Situationist pamphlet), xiv
Terry, Sonny, 170
Thatcher, Margaret, 87
Theater and Its Double, The (Artaud), 18,
234n15
theater of the absurd, 2, 5, 56
Third Mesa (Arizona), 1, 8, 22, 26, 28, 31
Thomas, "Tie Dye Annie," 134
Thompson, Dorothy, 60, 83
Thompson, Hunter S., 117, 118, 137
THX 1138 (film, dir. Lucas, 1971), 110
Tilly, Charles, 198
Tingen, Paul, 44
Tips for Zips (fanzine), 207
Toland, Gregg, 104, 105
Tomlin, Lily, 12, 202; *And That's the Truth*
(1972), 248n40; *Modern Scream* (1975),
201; *On Stage* (1977), 202
Tomlinson, Gary, 44–45
Too Close to Be Newsletter (fanzine), 207
Tosches, Nick, 118
Toubkin, Christina, xiv, xvfig.
Tower Records, 135
Townshend, Pete, 218, 248n34
Trailing Clouds of Glory (fanzine, 1974), xiv,
xvfig.
Treleaven, Harry, 150
Trips Festival (1966), 47
Triumph of the Will (film, dir. Riefenstahl),
86
Tubes, the, 203
Turner Fred, 47
turntablism (hip-hop practice), 215, 216, 217

Turtles, the (band), 47
TV Guide, 174, 195
TV Guide parodies, 165–67, 168fig.
TV or Not TV (Proctor and Bergman, 1973),
181, 183–85, 186, 189
"TV Set, The" (early draft of *Dwarf*), 95, 127,
246n16
2001: A Space Odyssey (film, dir. Kubrick,
1968), 139, 179

Uhelszki, Jaan, 202
UHF broadcast signal, 176, 181
Ultraman (Japanese TV show), 255n29
underground press, 4, 8, 19, 135; sex indus-
try and, 190–91; Spiritus Cheese and, 77
Under Milk Wood (Thomas), 24
United States, possibility of fascism in, 34,
86, 114, 245n129
United States of America (band), 7, 170
UNIX, 161, 252n30
Unwin, Stanley, 16
"Up Against the Wall Motherfuckers," 146
Uricchio, William, 164
Usenet groups (1990s), 15, 51, 212
Usher, Gary, 23, 24, 37, 39; postproduction
techniques and, 44; theatrical production
effects and, 38
USIA (United States Information Agency),
51, 53, 159

Vandenberg, Gerard, 6
van Doorn, Johnny "the Selfkicker," 6
Variety magazine, 96
Vellender, Phil, xiv
Verbivore zine, 207
Verma, Neil, 15, 58, 60, 240n43
VHF broadcast signal, 65, 181
video games, 257n11
Vienna Communist Youth Festival, 7
Vietnam War, 3, 18, 49, 57, 73, 82, 172;
African American casualties in, 63; allegory of, 50; antiwar movement, 122, 136;
antiwar stance of *Zachariah* film, 99;
Cambodia invasion, 91, 127, 137; computers used in, 158; democratic citizenship and, 62; escalation of, 22; FDR's
counterfactual "surrender" and, 84; Firesign Theatre fans among soldiers, 91,
112, 246n3; as "first television war," 53,
200; Gulf of Tonkin Resolution (1964),
85; "hearts and minds" propaganda campaign, 51, 85, 239n18; hi-fi equipment
used by US soldiers, 66; journalists' criti-

www.ingramcontent.com/pod-product-compliance
Lightning Source LLC
Chambersburg PA
CBHW020825270326
41928CB00006B/444